Living Landscapes

Heathland

First published in Great Britain in 2003
National Trust Enterprises Ltd
36 Queen Anne's Gate
London
SW1H 9AS
www.nationaltrust.org.uk

ISBN 0 7078 0348 9

Cataloguing in Publication Data is available from the British Library

Designed and art directed by Wildlife Art Ltd/www.wildlife-art.co.uk

All colour artwork by John Davis/Wildlife Art Ltd

Cover by Yellow Box Design. Original artwork by Alison Lang

Printed and bound in Italy by G. Canale & C. S.p.A.

Living Landscapes

Heathland

James Parry

THE NATIONAL TRUST

Contents

FACING PAGE Wavy hair-grass and bell heather at Dunwich Heath, Suffolk

Acknowledgements

One of the most revealing aspects of the research and writing of this book was how many heathland fans were prompted out of the closet. Clearly heaths have the power to move and fascinate, and it was heartening to find so many people willing to offer their time and support. Thanks go first to the many National Trust staff who helped with advice and information, particularly David Bullock and Matthew Oates at Cirencester, Julian Homer at Studland, Andrew Tuddenham in Pembrokeshire, and the many wardens and property managers at heathland sites across England, Wales and Northern Ireland. At English Nature, Steve Clarke, Rick Keymer and Isabel Alonso all provided helpful suggestions and background. I am also grateful to Andy Fale, Reg Land, Janice Kerby and John Tucker of the wildlife trusts for Dorset, Norfolk, Nottinghamshire and Shropshire respectively, to Tom Williamson of the Centre of East Anglian Studies, University of East Anglia, Paul Barwick and Nick Gibbons of Forest Enterprise, Piers Chantry of Defence Estates (Ministry of Defence), Jude James of the New Forest Museum, Oliver Bone of the Ancient House Museum, Thetford and Gareth Beech at the Museum of Welsh Life. Chris Howkins was an invaluable source of information on heathland products and John Box likewise on heathland's industrial past. Fiona Dickson provided advice on heathland bee-keeping, and John and Stanley Beynon of Pembrokeshire helped reveal some of the history of their local heaths. Thanks go also to Liz Dittner for her editorial expertise and to Fiona Screen of National Trust Publications for her help with picture research and for steering the book through to completion. Finally, many thanks to my partner Andrés Hernández for putting up so patiently with innumerable (and compulsory) expeditions to obscure tracts of heathland.

Foreword

One of the best ways to enjoy a late summer's day is to find yourself a patch of heathland, lie down and close your eyes. Not only will you feel the pulsating warmth of the heath, but the humming of a myriad insect wings will surround you. A few up close at first and then, as you sink into the heather, the whirring of tens, hundreds, thousands of tiny pairs of wings will fill the air. The heath will be alive, full of flying insects going about their business. Many of these will be bees, harvesting the seasonal glut of heather nectar from the purple bell-shaped flowers. Open your eyes and you will be greeted by a floral spectacle every bit as spectacular as it is rare.

And that is the pity. Heathland is one of our most precious habitats, but one that is caught in paradox. A habitat largely created by man and intertwined closely with our own history, we spent much of the twentieth century wrecking it. Unappreciative of its true beauty and riches, we tore it up, ploughed it, fertilised it or encased it in concrete. Instead, we should have been shouting about it from the rooftops, proud that we had created such a diverse and wonderful place.

The following chapters by James Parry explore and celebrate this special world, revealing the unique role played by heathlands in our social and natural history. Heathland converts have long known this, but now is the time to spread the word further. I encourage you, urge you, to get out there and visit a heath. Maybe you too will fall under the spell of the stone curlew's stare, the nightjar's 'music' and the hum and bustle of the heather and its harvesters.

Nick Baker
Broadcaster and television presenter

Chapter One

❖

Heaths in context

HEATHS IN CONTEXT

A Lost World

> ... he who has once vibrated with the thrill of the heathland is never quite the same again ... It fascinates but few, but those who have once come under its spell are ever after its slaves.
>
> W.G. Clarke (1925)

When these words were written by William Clarke, an outstanding Norfolk historian and naturalist, the heaths he knew and loved were disappearing fast. The East Anglian Brecks, a huge expanse of unspoilt heathland which Clarke understood and appreciated perhaps more than anyone else, were falling victim to afforestation, road-building, housing and agricultural expansion. His book, *In Breckland Wilds*, remains a beautiful and moving eulogy to a world virtually lost today. Only fragments of the great Norfolk and Suffolk heaths survive, either marooned between fields of sugar beet, housing estates and by-passes or found clinging, limpet-like, to the open rides of ancient landscape that survive between vast plantations of fir and pine. Aflame with purple heather in late summer, these remnants flicker like the last symbolic embers of a once roaring fire. Across the rest of Britain the picture is equally dramatic: some 85 per cent of the heathland extant in 1800 has since been destroyed, much of it in the last hundred years.

Clarke's acknowledgement that the appeal of the heath was rather limited at least partly explains its general demise. During most of the twentieth century heathland had few champions and many detractors, being regarded widely as a fair target for destructive attention at a time when land values were usually rising and the pressure on open space becoming ever more concentrated. Heathland 'wastes' were perceived as just that: an unnecessary and barren feature in a landscape where the emphasis was increasingly on usefulness and productivity. Yet, in the rush to sweep them away under a sea of tarmac, golf courses, houses and intensive crops, a whole world of fascinating social history and immense ecological value was lost. Far from being wasteland, Britain's heaths are testament to the industry and ingenuity of our forebears and to a way of life – now vanished forever – that was forged on the very margins of society: the world of the furze-hacker, broom-squire and turf-cutter. The heaths are also home to a specialised range of wildlife which, in the case of many species, is highly restricted and not found widely in other British habitats. Only now, at the eleventh hour, have we come to realise quite what we were allowing to disappear.

PREVIOUS PAGE An autumnal scene in the New Forest. Despite its name, the Forest contains extensive tracts of open heath and the exploitation of these was an integral part of the area's traditional management.

Lowland heath is now recognised as one of Britain's most important landscapes in terms of wildlife conservation, and is a high UK Biodiversity Action Plan priority. Virtually all areas of lowland heath now come under one or another of the various statutory designations (over 90 per cent are designated as Sites of Special Scientific Interest, for example, although this does not by itself guarantee either their protection or sympathetic owners) and very considerable resources are now being devoted to heath restoration and improved management. Heathland has become fashionable.

This engraving from the mid-nineteenth century shows the steppe-like landscape of Thetford Warren on the Breckland heaths, home at that time to stone curlews and lapwings, but now mostly lost to forestry and agriculture. The outline of the warren lodge can just be seen on the horizon (see also p.76).

What is heathland?

The heathlands that we see today are, with a few notable exceptions, essentially manufactured landscapes, the result of human activity. Although the precise origins of early heathland remain unclear and are still the focus of lively discussion (see Chapter Two), it is likely that much of it was originally wooded to a greater or lesser degree. Cleared by man for use as timber and fuel, and to make way for rudimentary agriculture and animal husbandry, the resulting landscape was then managed, usually by grazing stock. This both prevented the regeneration of the forest and provided an economic return. Exceptions to this pattern include the maritime heaths of the coastal fringes of western Britain, where extensive tree cover was never common, but there may well be others, their evolution less well understood. Even so, heathland can be regarded as a transition landscape, i.e. between woodland and woodland. Without continued intervention by man, most of today's heaths would revert to forest. Meanwhile, heathland has a particular vegetation and characteristics of its own, and provides appropriate conditions for a range of specialist wildlife.

Exactly what constitutes 'heathland' continues to be a source of some debate, and although there are several different types of heath, they all have defining qualities in common. The single most obvious characteristic about heathland is the low incidence of trees. Heathland is a predominantly open and flat landscape in which dwarf shrubs predominate, usually heathers (*Calluna vulgaris* or *Erica* spp.) and gorse (*Ulex* spp.).

Cavenham Heath in Suffolk is a classic Breckland heath, with extensive areas of mixed grasses and ling. Regular management is required to prevent the encroachment of birch scrub.

Indeed, the origins of the word 'heath' betray this association. It stems from the Old English *hæð* and from this probably also comes the word 'heather'. Heath is currently defined by the Oxford English Dictionary as 'open, uncultivated ground; a bare … flat tract of land … covered with low herbage and dwarf shrubs, especially heather or ling'. This is helpful as far as it goes, but most heaths do, in fact, contain at least isolated stands of scrub and trees – often birch or Scots pine. The rate to which the spread of these trees is controlled is a major determinant in the quality and character, if not the continued existence, of the heath itself. Furthermore, certain species of grass are also prevalent on heathland, and in some cases – most notably in the Anglian Brecks – may constitute the dominant vegetation.

The importance of soil

The variations in different types of heath are determined by the underlying soil types and general terrain, as well as by climate. Soil is a key defining characteristic of heathland in the sense that heaths are usually found on nutrient-poor, acidic soils that have developed on freely draining sands and gravels. The easy drainage and general absence of tree cover on heathland gives rise to heavily leached soils in which nutrients are washed out more rapidly than they can be replenished by either the vegetation they support or by the natural upward process of evaporation. Plant growth is, in any case, inhibited by the low concentrations of key nutrients, such as phosphorus and nitrogen. This results in the occurrence of much heathland on podsols, bleached and acidic strata of soil matter, the upper levels of which are usually the colour of light wood ash. Indeed, the term *podsol* is derived from the Russian words for 'under' and 'ash'. The ash-coloured layers sit below the surface accumulation of decomposing organic matter, which is often rather scant and slow-acting in the relatively cool climate and highly acidic conditions typical of most heathlands. The humic acid produced by decomposition is dissolved by rainfall and washed through into the mineral material below, activating iron oxides en route and then leaching these down into an impervious layer of iron compounds known as a pan. Few plants can penetrate this pan and so, for obvious reasons, most heathland vegetation is very shallow rooted. Indeed, ecologist C.H. Gimingham (1975) cites a Dorset heath where samples revealed 92 per cent of all plant roots to be confined to the top twenty centimetres (eight inches) of soil.

However, heathland is not restricted to acidic soils. There are numerous examples of 'chalk heath', areas of calcareous soil that support classic heathland flora such as heather alongside more lime-loving species like salad burnet and dropwort. Such cases usually occur where small but discrete areas of acid soil sit over chalk or limestone, but they serve as a timely reminder that the picture is complex and that heathland will readily arise as a response to what are often very localised soil conditions, so long as the appropriate forms of land use are in place.

This cross-section shows the typical stratification of sandy heathland soils. Note the layer of heavily leached ash-grey soil immediately beneath the heather mat and topsoil, and the dark iron 'pan' below. This pan is virtually impenetrable and serves as a potent obstacle to any plant with deep roots. Underneath the pan is an orange-brown layer of iron-enriched material, with parent sandstones and gravels below.

Types of heathland

Britain's National Vegetation Classification lists up to fourteen distinct categories of lowland heath community, which can be simplified into seven general types:

i) dry heath, the most familiar type of heathland, found on well-drained soils and typically dominated by heather species (mainly *Calluna*), with areas of gorse and grasses;
ii) lichen heath, communities of lichen species living on bare, sandy soils;

iii) grass heath, often found on calcareous soils, dominated by grass species such as sheep's fescue, but often containing patches of heather and/or gorse;
iv) humid heath, where soil drainage is impaired to some extent and damp patches occur in all conditions. Heathers dominate, often with purple moor-grass;
v) wet heath, where drainage is markedly poor and waterlogged areas persist even in times of drought. Certain heather species dominate, along with purple moor-grass, sundews and bog mosses such as *Sphagnum*;
vi) valley mire or bog, occurring when low-lying valleys intersect heathland and usually found on peat. Very wet, dominated by *Sphagnum* and also featuring species such as cotton grass and bog asphodel;
vii) maritime and coastal heaths, both usually dominated by heather species. Maritime heath is generally found on clifftops and often contains typical coastal species such as thrift; other coastal heaths occur on dune systems, where the heather is mixed with grass species such as sand sedge.

Maritime heath at St David's Head in Pembrokeshire. The vegetation on the exposed seaward slopes is kept short by the effects of wind and salt, but further inland the heath is taller and more diverse.

Part of Frensham Common in Surrey. Owned by the National Trust, this is an outstanding and varied area of heathland, supporting a rich variety of wildlife.

Ling is the most successful British heather species and capable of thriving in all sorts of unpromising conditions. Its ability to do so may help explain its historical reputation for being lucky.

Heathland is rarely composed solely of one community and many of the larger sites will feature several of the above within their overall extent. Indeed, any heathy landscape is likely to contain a number of smaller, more specific, habitats which individually support a range of different flora and which collectively comprise the whole heathland plant community. These habitats can range from clumps of mixed scrub, stands of gorse and areas of ericaceous cover, through stretches of open ground and grassland to poorly drained wet heath. For example, the heathland known to William Clarke in East Anglia was mainly open grassy heath featuring hardly any trees and interspersed with patches of lichen, clumps of heather, bare expanses of sand and even areas of open water (the mysterious Breckland meres, which empty and refill with water regardless of immediate rainfall patterns). It is precisely such mosaics of habitat that make heathland so valuable to wildlife. The common denominator throughout, however, is that any vegetation (and associated fauna) must be tolerant of the generally impoverished conditions that apply in the lowland heath landscape.

The heather community

> The ecology of *Calluna* is, to a large extent, the ecology of the European heathlands.
>
> Nigel Webb, *Heathland* (1986)

As already outlined, heathland is commonly defined and characterised by the presence and, usually, the domination of members of the heather family, *Ericaceae*. Heather or ling, *Calluna vulgaris*, is by far the most common and widespread member in Britain and Ireland. It is equally widely distributed in Europe, from the Azores to beyond the Ural

Mountains in Russia (although heathland is much more restricted in range), and will grow readily in a variety of locations from sea level to over 1,000 metres (3,281 feet). Yet throughout this extensive range its form and character are remarkably constant, with hardly any regional variation, although several local forms are known. Essentially a low-growing and much-branched shrub, it rarely exceeds 125 centimetres (49 inches) in height and on average attains 60–80 centimetres (24–32 inches) or so. The woody stems branch from the base of the plant, often extending laterally close to the ground and forming a fairly dense understorey.

Leaves are small and grow along the stem in opposites. The small and compact flowerheads cluster together towards the top of the stem and occur in colours ranging from white to deep purple, with pale to medium pink being the most common. Flowering takes place during August and September, and can make for a dramatic display when at its peak. Seeds are shed – in vast quantities – during autumn, mainly with the help of the wind. Depending on conditions, they can germinate after eight weeks or so.

Westleton Heath in Suffolk, showing ling in various stages of growth. The structure of the dead plants in the right foreground has fallen apart, thereby promoting the germination of dormant seeds in the ground litter below.

However, adequate levels of moisture are required for germination to take place, and if the soil is too dry the seeds will lie dormant until things improve. Seeds can remain inactive for sixty or more years, a surprising but decidedly useful quality in view of recent efforts to rescue and restore heathland previously put under forestry, for example (see p.131). The high levels of seed production create a vast seedbank in the litter beneath the plants, although failure rates and mortality are quite high.

Although widespread across Britain, bell heather is less robust than ling and so generally gives way to its more vigorous relative. However, in certain drier situations its ability to withstand parched conditions can give it the edge. A male Montagu's harrier is passing low overhead.

The lifespan of an individual ling plant can be anything up to forty years. During its overall growth period the plant demonstrates various forms and habits which, broadly speaking, constitute four distinct phases. The first is the pioneer phase, the time during which the seedling establishes itself and develops into a small bush. This usually takes five or six years, and is followed by the building phase, lasting ten years or so, during which the bush fills out and becomes better established. It then enters the mature phase, lasting a further ten years, when growth becomes progressively less vigorous but seed production is at a maximum. Towards the end of this phase the centre of the bush tends to open out and fall apart, marking the advent of the final phase, the degenerate or veteran. By now growth has slowed down completely and the plant structure collapses, exposing the bare ground below. In most cases, death follows after a few years. Meanwhile, however, new seedlings will be establishing themselves in the open areas that have been created by the older plant's demise. However, it is important to note that, thanks to factors such as historical grazing patterns and, indeed, contemporary heathland management practices, not all individual heather plants will follow quite such a schematic pattern. In addition, the duration of each phase varies according to location and climate.

Ecologically, ling is more flexible than other types of heather. It shows greater tolerance of marginal soil and moisture conditions, and so is able to thrive in a greater variety of physical locations. This flexibility helps explain why, in most conditions, it generally demonstrates higher levels of germination and seedling survival than either of the other two common species of heather, bell heather (*Erica cinerea*) and cross-leaved heath (*Erica tetralix*), and why, therefore, it is the most dominant and successful member of the heather family in Britain.

Bell heather is as widely distributed in Britain as ling, but in most localities it is much less pervasive and does not normally form the extensive tracts typical of its more robust relative. It has a more compact habit, rarely exceeding 60 centimetres (24 inches) in height, and is markedly less woody in appearance; consequently, when growing alongside ling – which it often does – it tends to be dominated by its neighbour. The flowers are larger and a deeper pink than in ling (or in cross-leaved heath, for that matter) and the species derives its common name from their bell-like shape. The leaves – as with all members of the *Erica* genus – grow in whorls at regular intervals along the stem. Furthermore, bell heather flowers from July, slightly earlier than ling, and tends to be restricted to drier heathland areas; wetter soils inhibit both its germination and seedling development.

By contrast, cross-leaved heath is very much at home on both humid and wet heaths, where it can compete successfully with ling and, in some cases, even form the dominant vegetation. It grows extensively across Britain and Ireland and is easily identified by its characteristic foliage colour – much paler and greyer than either ling or bell heather – and by its flesh-pink flowers, as well as the characteristic whorls of four leaves arranged like crosses down its stem.

Cross-leaved heath can tolerate much wetter conditions than either ling or bell heather, and thoroughly justifies its other name of bog heather. However, all three species will grow alongside each other in the right terrain, with ling and bell heather on drier mounds or tussocks and cross-leaved heath in the damp hollows between.

Finally, there are two further species of heather worthy of special note: Cornish heath (*Erica vagans*) and Dorset heath (*Erica ciliaris*). Both are highly restricted in terms of their distribution, although they can be abundant in certain favoured localities. In Britain and Ireland the former is found only in County Fermanagh and on the Lizard, where it grows in stands up to 80 centimetres (32 inches) high and has developed at least six colour variants. Dorset heath, characterised by its densely bunched growths and striking flower spikes, grows extensively in the county from which it derives its name, mainly on the Isle of Purbeck, as well as in parts of Devon and Cornwall.

The main point about the heather community is that although individual species may vary in terms of their specific requirements, collectively they are highly successful partners. Some heathlands in southern England contain up to four different species of heather, on a patchwork quilt of different degrees of abundance reflecting subtle

variations in soil and local conditions. The importance of this partnership is paramount, as on most heaths the heathers form the framework for the entire heathland plant community, which is described in more detail in Chapter Four.

Heath or moor?

In Britain, the words 'heath' and 'moor' have traditionally been confusingly interchangeable, but this relaxed use of terminology masks a fundamental difference behind the apparent similarity of terrain and the prevalence in both landscapes of ericaceous vegetation. Landscape historian Oliver Rackham, among others, has defined this difference as being one of soil type: moorland sits on peaty soils and heathland does not. Although an oversimplification, this differentiation is essentially correct and has been refined to a useful general understanding of 'lowland heath' – with which this book is concerned – to be that confined to land below 300 metres (985 feet) in altitude. Wet, peaty soils are most common above this level, and dry, sandy ones generally restricted to land below it, hence the distinction that can be made between moors (in the uplands) and heaths (in the lowlands). However, peat deposits do occur on lowland heath, most notably when low-lying valleys transect an otherwise 'dry' heath. Where this happens bogs or so-called valley mires can arise, which, although supporting a flora more akin to that found in the wetter conditions more typical of higher elevations, are best considered as a special, but integral, feature of the lowland heath landscape.

Permanently wet bogs and areas of open water are an integral part of the heathland landscape in many areas. They constitute a rich habitat for wildlife, especially plants and invertebrates. This example is at Povington in Dorset.

As if the definition of heathland were not uncertain enough, the picture is further obscured by the very loose historical attribution of the word 'heath' to any type of supposedly barren, open ground. In some cases this could mean simply areas of alkaline grassland with few real heathland characteristics as we now understand them. However, the use of such terminology is important because of the light it sheds on historical and cultural attitudes to heathland and landscape generally.

One final point on nomenclature: the word 'heath' can also cause misunderstanding due to its application not only to a type of landscape, but also to some species of the *Erica* genus, most notably the cross-leaved heath, Cornish heath and Dorset heath. Such confusion is perfectly logical, however; the word 'heather' has its root in the Old English hæð, which was used to mean both the place and the plant: evidence of just how synonymous the two are.

Britain's remaining heathland is highly scattered and disjointed. For the sake of clarity, the surviving fragments are shown on this map at a substantially larger scale than reality. Collectively this land amounts to approximately the same area as that covered by the Isle of Wight.

Heathland distribution

The distribution of lowland heath is essentially defined by the soil types described above and by climate. The right combination of the two creates the conditions favoured by ericaceous vegetation, notably ling. Ling grows most enthusiastically in moist, temperate oceanic conditions, such as those found along the Atlantic seaboard in north-western Europe including, most notably, the British Isles. Plentiful rainfall, generally high humidity and a lack of extremes of temperature characterise this zone; monthly mean temperatures range between 0° and 22°C (32°F and 72°F), with average annual rainfall of 600–1100mm (24–43 inches). The headquarters of European lowland heath is firmly based around the North Sea littoral, where these conditions apply most typically – in southern and eastern England, Belgium and the Netherlands, northern Germany and then north into Denmark, southern Norway and south-western Sweden. However, lowland heathland occurs – albeit more sporadically and with greater variation – across a far wider area than this, one that includes all of Britain and Ireland, much of Scandinavia, northern France and the north of Spain and Portugal.

The climate of some of these regions, both in terms of temperature and rainfall, lies outside the ideal conditions for ling, but heathland occurs nonetheless, even if it may be less well developed. In areas of higher rainfall, for example, lowland heath may be restricted to the most well-drained soils (such as those on slopes), whilst conversely in drier and hotter contexts it may occur primarily in coastal locations where temperatures are more moderate and humidity higher. Whilst ericaceous plants generally favour moist and cool conditions, their tough waxy leaves do enable them to minimise water loss and thus tolerate remarkably well both heat and lengthy dry spells. This is important on heaths that have developed in conditions slightly outside the normal range of a temperate oceanic climate, such as those in the Norfolk and Suffolk Brecks, for example. Inland East Anglia has a more continental climate than the rest of Britain, with hotter summers, colder winters and considerably less average annual rainfall. Frost is known in all months and temperatures can descend to –15°C (5°F) or lower in winter and exceed 30°C (86°F) in summer. The low winter levels may explain why ling, which is vulnerable to severe frosts and can survive prolonged low temperatures only if insulated by a cover of snow, is often a less dominant heathland vegetation here than in other parts of Britain and Ireland.

Wherever it may grow, typical heathland vegetation is tough and resilient. In particular, it shows a marked resistance to wind, which helps explain its success in coastal areas such as Cornwall, Pembrokeshire and Brittany. In areas such as this, where salt spray can also be a factor, clumps of heather are often sculpted into crouching, undulated forms, often dead or dying back on the seaward side but thriving and extending wherever a little shelter is offered – usually to the lee. Equally, both heather and gorse show remarkable powers of recovery after fire and can regenerate successfully from charred stumps (see p.123).

The extent of European heathland is now dramatically reduced from its peak in the nineteenth century, when it is estimated that there was some 30,000km^2 (11,580 square miles) of lowland heath in north-western Europe (including Britain). Now only 4,000km^2 (1,544 square miles) or so survive, a decrease of 86 per cent since the nineteenth century. Some countries, such as Germany and the Netherlands, have lost as much as 95 per cent of their heathland in little over a hundred years. Economic development, particularly during the latter half of the twentieth century, forced the abandonment of traditional heathland practices and encouraged the destruction of what had been a major landscape component in rural life across much of western Europe. Much of the heathland that remains is dangerously fragmented and seriously degraded through neglect and/or misuse.

Carn Galver in West Penwith, Cornwall. The heathland in this area is among the best in the county and contains a rich archaeological heritage.

British heathlands

In Britain the story of heathland decline and disappearance is almost as depressing and as comprehensive as in Continental Europe. However, England in particular retains a major proportion of surviving European lowland heath and the United Kingdom clearly has a global responsibility to protect and enhance what little is left. There are currently some 750–1,000 heathland sites in the United Kingdom, covering over 58,000 hectares (143,500 acres). These are distributed widely – but not consistently – across southern England from Cornwall to Norfolk, with outposts in west Wales, the Midlands, Yorkshire, Cumbria, Northern Ireland and Scotland. Most surviving heaths are shrunken vestiges of their former selves, and fragmentation remains a major issue in heathland conservation. Indeed, evidence suggests that in the past areas of heath were commonplace in almost all counties; for example, the small patches of heather that continue to survive in the beechwoods of the Chiltern Hills in Buckinghamshire are doubtless relicts from a time when heathland was much more widespread there.

The National Trust's Witley Common in Surrey is a mixture of heath, scrub and woodland edge. Evidence of the area's exploitation by man goes back thousands of years and ranges from Bronze Age burial sites to the remains of Second World War army camps.

In the south west of England, in Cornwall, the maritime heaths of the Lizard peninsula and West Penwith are among the finest of their type. These probably constitute the main sites in Britain for heathland as a climax vegetation: the prevalence of strong winds and salt simply prevents tree growth in these locations, and so heather and gorse are at the top of the pyramid. The Lizard in particular is noted for its rare plantlife, and further areas of good-quality heath survive on the north Cornish coast around Chapel Porth. In neighbouring Devon, Dartmoor and Exmoor are both largely covered in heather, gorse and other classic heathland vegetation, but as they both lie mostly above 300 metres (985 feet) they cannot be classified as lowland heath. However, similar conditions occur at lower levels in the east of the county, where there are important areas of mixed dry and wet heath. Aylesbeare Common, for example, managed by the Royal Society for the Protection of Birds, is a particularly valuable site for wildlife.

Dorset is one of the headquarters of British heathland. The area of the Poole Basin contains some of the most important stretches of heath in Britain, sadly just remnants of the vast tracts that once covered much of the county and which were described so evocatively by Thomas Hardy as his fictional 'Egdon Heath'. Home to significant reptile populations, notable birds including Dartford warbler, and important areas of the scarce and localised Dorset heath, the best surviving examples include Studland and Godlingston Heaths and Hartland Moor. However, the single largest extent of lowland heath in Britain, covering some 15,000 hectares (37,000 acres), is actually in Hampshire's New Forest. Not only are the Forest heaths highly significant for wildlife, but they are also of immense cultural importance in terms of their social history and system of traditional land use rights. Various types of heath are represented here, including extensive areas of wet heath and mire as well as some of best stands of mature gorse anywhere in the country. In addition, on the Greensand ridge which runs through the north of the county there are some important areas of heath, such as Bramshott Common, Ludshott Common and Woolmer Forest, the last site one of a handful that are home to all twelve species of British reptile and amphibian.

The Greensand deposits continue eastwards into Surrey and support areas of heathland for much of their length. The most important sites are located around Frensham, Hindhead and Thursley, and constitute a surprising survival in what is otherwise a densely populated region: arguably the last wilderness in south-east England? Mostly comprising dry heath but with important areas of wet heath too, they are all rich in wildlife and Thursley National Nature Reserve alone is home to more than 10,000 species of insect. Over the county border in Sussex, there are important heaths on the north side of the South Downs, most notably at Iping Common, and also in Ashdown Forest. This was once part of a royal hunting forest and, like the New Forest, was traditionally managed by the application of common rights (see p.44). The result is an interesting mosaic of different heathland vegetation types.

Cannock Chase, Staffordshire was established as a royal forest during Norman times. Most of the trees were subsequently felled to provide charcoal for the local iron-working industry, resulting in large tracts of heathland.

East Anglia also has notable areas of heathland. In the Brecks of south-west Norfolk and north-west Suffolk are the fragments of what was formerly an extensive range of predominantly grassy heath with a highly specialised flora and unusual birdlife, now mostly lost to forestry and agriculture. The Norfolk Wildlife Trust's Weeting Heath and English Nature's Cavenham Heath, just over the county boundary in Suffolk, are both important survivals. There is also a chain of remnant heaths on the north Norfolk coast, centred around Salthouse, Kelling and West Runton, as well as a few scattered sites of predominantly wet heath and valley mire in west Norfolk, especially at Dersingham Bog. In coastal Suffolk, there is an outstanding area of heathland known as the Sandlings, once an area of extensive sheep walks but now much fragmented. The National Trust's Dunwich Heath is typical of what remains.

Two Breckland specialities: spiked speedwell and stone curlew. The twentieth-century destruction of much of the area's heathland led to the decline and near-extinction of both species.

The former heaths of the East Midlands are now greatly depleted, reduced to only four per cent of their extent in the early nineteenth century. Remnants still survive in Lincolnshire, where extensive grassy heath similar to that found in the Anglian Brecks once existed on the Coversands in the north of the county. Meanwhile, the district just south of Lincoln is still known as 'the Heath' today, even though any heathland has long gone in the face of intensive agriculture. Fragmented patches of heath are also found in Leicestershire (particularly in Charnwood Forest) and in Nottinghamshire, where the celebrated Sherwood Forest was formerly comprised substantially of heathland. An ambitious plan is now underway to partly restore the forest and its heaths, currently an unappetising mixture of arable farmland, coniferous plantations and semi-industrial wasteland.

Across the West Midlands there are also various relict heaths, mainly in Cheshire, Shropshire, Staffordshire and Worcestershire. Many of these sites became heathland as a result of extensive tree-felling in support of industrial development during the eighteenth and nineteenth centuries and, until recent decades, were maintained as heathland by the grazing of sheep and cattle. Many have direct evidence of an industrial past, such as disused mines and old quarries, and some are now located on the urban fringe, which brings a whole range of particular pressures, threats and opportunities (see Chapter Five). Meanwhile, at sites such as Cannock Chase, the presence of certain plant species more typical of upland moors, such as crowberry, indicates the transition that these heaths represent between the classic lowland heaths of southern England and the moors of northern Britain.

In the north of England most dwarf shrub communities occur at elevations higher than that required to qualify as lowland heath, but there are a few notable exceptions in Cheshire (notably Thurstaston Common on the Wirral), Cumbria (areas of dune heath at Sandscale Haws, for example), Lancashire (the dune heath at Ainsdale, as well as other fragments such as at Heysham Head) and in the Vale of York. In Wales the best lowland heaths are located in Pembrokeshire, mostly on St David's Head, and near Swansea on the Gower peninsula. Both these areas are very rich botanically. Further north, in Gwynedd, there are good examples of dune heath near Harlech and some areas of maritime heath on both Anglesey and the Llŷn peninsula. The Welsh coastal heaths are particularly significant in terms of their value to the rarest native member of the crow family, the chough (see p.129). Meanwhile, the topography of Scotland is such that lowland heath is mostly restricted to coastal areas, most notably in Argyll, Caithness, Orkney and along the east coast, such as on the Sands of Forvie close to Aberdeen. Equally, there is very little true lowland heath in Northern Ireland; the best example is the excellent dune heath complex at Murlough in County Down.

A bleak winter scene on Mannington Heath in Dorset. Traditionally regarded as a hostile and unforgiving landscape, heathland never appears more so than during the colder months of the year.

The View of the Heath

Historically, one's view of heathland was generally determined by vantage point. For those living on or near it, the heath could provide both a home and a livelihood. But to outsiders and visitors it was a landscape mostly regarded with suspicion, if not downright contempt – an attitude which persisted in some quarters until surprisingly recently. Much of this negative perception can be attributed to the bleak and desolate quality so characteristic of heathland. For much of history, such overwhelming emptiness was regarded as eerie and threatening, fit only for nefarious deeds and even sorcery. Certainly, the heathland landscape was regarded by William Shakespeare as an ideal setting for much of the scheming and plotting so characteristic of his tragedies. One need look no further than the witches on the 'blasted heath' of *Macbeth*; Shakespeare had never visited Scotland, and his image of the moorland landscape there was doubtless based in part on heathland closer to his

This early twentieth-century postcard depicts the Devil's Jumps, a series of hills overlooking the heaths of Hindhead. Such imagery helped reinforce the prevailing idea of heathland being god-forsaken and fraught with danger, both physical and spiritual.

home in the English Midlands. Conditions on Shakespeare's fictional heaths were generally sinister, the weather invariably stormy. And those unfortunate enough to live there were to be pitied. In *King Lear*, the Earl of Kent, upon arriving on a heath in bad weather, was astonished to find signs of habitation (albeit an abandoned hovel):

> Poor naked wretches, whereso'er you are,
> That bide the pelting of this pitiless storm
> How shall your houseless heads, and unfed sides,
> Your loop'd and window'd raggedness, defend you
> From seasons such as these?

This idea of heaths being some sort of cheerless twilight zone on the very edge of civilisation held currency for a long time. Heaths were places where ghosts and ghouls held sway, where decent behaviour and everyday values were suspended and where unspeakable acts and crimes were endemic. Some heaths were the notorious haunt of footpads and highwaymen, and certainly many had more than their fair share of gibbets; it is no surprise, therefore, that in many eyes heathland was a landscape of fear and foreboding. Nowhere did this view hold greater currency than in the Devil's Punch Bowl at Hindhead in Surrey, described by radical writer and journalist William Cobbett in his *Rural Rides* (1830), an account of a journey he made across southern England from 1821–26, as 'the most villainous spot that God ever made'. Much of this fearsome reputation stemmed from an infamous murder that took place there in 1786, when three ruffians attacked a solitary sailor travelling to Portsmouth on the isolated road along the rim of the Punch Bowl. The culprits were captured shortly thereafter and hung from a gibbet nearby. However, this incident took on a magnitude out of all proportion to its real import and became a powerful and enduring symbol of the dangers of the lonely heath. Hindhead and the Punch Bowl have taken a long time to recover; the squatters who traditionally lived in the Bowl were regarded by those outside as little better than common criminals, a smear that lingered until quite recently.

Indeed, throughout much of history heathland dwellers were considered to be very much at the bottom of the social pile, virtual outcasts, living on marginal land that no-one else could be bothered to deal with. As late as 1793 Hounslow Heath and Finchley Common, both near London (and now engulfed by suburbia), were described as being fit only for 'Cherokees and savages'. This cultural prejudice goes back many hundreds of years and, as historical ecologist Chris Howkins argues, is almost certainly related to the ancient idea that heath-dwellers were pagans. The Latin term *pagus* was originally used by the Romans to mean village or settlement, and from this came *paganus*, used to denote the inhabitant of a village or, more generally, a country-dweller; the word came to have negative connotations, being used in unfavourable comparisons with the superior strata of society, such as trained soldiers and 'citizens' in the original urban sense. With the coming of Christianity to the Roman Empire in the fourth century AD, those who lived on the remote heaths were among the last to be 'enlightened' and to remain *pagani*. The Scandinavian and German tribes associated these peoples closely with the heather amongst which they lived, *der Heide* in German, and from this term it is a relatively straightforward step to *die Heiden*, meaning pagans, from which we get the English word 'heathen'.

However, there is a discernible difference between the general history of suspicion and antipathy towards heathland and its inhabitants and the later, more overt, desire to destroy the heath and somehow 'civilise' the people that lived there. The key point here is that for many centuries heathland was considered an integral part of the working landscape. It often formed part of the manorial waste attached to a feudal property and was usually subject to defined rights of ownership and usage. Much heathland was common land, which is why it was so vital to the poorer members of society, who were unable either to acquire land or gain adequate access to it or its products in any other legal way. While the feudal system survived, heathland was not in any great threat. Indeed, its continued existence was positively encouraged by the integrated land use that feudalism entailed; regular cycles within each manorial demesne of tree-felling, grazing, burning etc. However, as man's ability to control his environment more comprehensively was fuelled by agricultural improvement and technical advance, so heathland was increasingly derided by the influential advocates of so-called progress. These new thinkers regarded heathland as barren and unproductive, an obstacle to development, a 'waste' in more than the original sense of the word (for which see p.43). This view was already apparent in the writings of Daniel Defoe who, during his travels through Britain in the 1720s, noted the heaths of the New Forest as being 'undoubtedly good and capable of improvement', and became especially prevalent during the late eighteenth and early

Children collecting furze faggots on a Dorset heath, late 1890s. Life on the heath was tough, and all members of the family were called upon to pull their weight.

nineteenth centuries, as the Industrial Revolution gained momentum. The conversion of heathland to productive arable land was regarded as a laudable objective, the failure to use such land profitably being nothing less than unthinkable and morally reprehensible in the modern age. Meanwhile, the advent of Parliamentary enclosure marked another nail in heathland's coffin. By enclosing large areas of common land, much of it heath, rapacious landowners stripped many heathlanders of their homes, their rights and their livelihood. Opposition by those so dispossessed was taken as further evidence of the contrary and obstructive nature of these marginalised folk. The only value of heathland was as farmland; it was of no value *per se*.

Yet heathland was not entirely without its supporters. The development of the Picturesque school from the late eighteenth century onwards, a reaction to the contrived and manufactured landscapes promoted by designers such as 'Capability' Brown, encouraged greater awareness of the natural beauty of the British landscape. Meanwhile, improved road communications were opening up hitherto remote areas to greater scrutiny and, potentially at least, to better understanding. Even so, Defoe's view of the heathland around Bagshot was typical of his time:

> Those that despise Scotland and the northern part of England, for being full of waste and barren land, may take a view of this part of Surrey, and look upon it as a foil to the beauty of the rest of England …

The Gravel Diggers by George Morland. Heaths were productive places, whatever their opponents claimed, and Morland's painting highlights the value to heath-dwellers of activities such as mineral extraction.

A century later, however, the view of artists at least had largely changed. By now (1815–25) the scope offered by heathland to offer exciting images of natural beauty had caught the eye of artists such as John Crome and John Sell Cotman, who both depicted Mousehold Heath near Norwich, and John Constable, who painted several views of London's Hampstead Heath. Meanwhile, painters such as George Morland and, much later, Sir Walter Russell preferred to depict the heaths as working places. Their respective works *The Gravel Diggers* (c.1792) and *Carting Sand* (1910) helped reinforce the notion – often lost in the rush to expand agriculture – that heaths could still be productive and worthwhile in the Industrial Age. This interest in heathland as a fruitful landscape was mirrored by trends on the Continent, especially in Denmark, Germany and Portugal, where mellow heathland scenes populated by worthy and contented workers became a favourite, if heavily idealised, subject for some artists. Nonetheless, there were still those who viewed heathland as the symbol of a certain poverty in the landscape rather than as a scene of wholesome beauty or source of inspiration. William Cobbett, for example, saw the desolate heath as the unhappy home of the impoverished agricultural workers whose condition he worked so passionately to improve. Certainly, he had no time for the romanticisation of what he saw as desperately downtrodden rural labourers at the mercy of exploitative landowners. For Cobbett, agricultural improvement – and the destruction of heathland in favour of productive arable land – might be a means of salvation for the rural poor if only it were properly managed.

Meanwhile, poets had begun to pay greater attention to heathland. Some castigated their predecessors for an approach that was too often emblematic rather than descriptive, and which therefore missed the real artistic value of such landscapes. Foremost among those interested in the detail of heathland was the poet John Clare (1793–1864), who portrayed the apparently bleak winter character of the heath in an enlivening way in his quirky yet evocative poem *Emmonsails Heath in Winter*:

> I love to see the old heaths withered brake
> Mingle its crimpled leaves with furze and ling
> While the old Heron from the lonely lake
> Starts slow and flaps its melancholly wing,
> An oddling crow in idle motions swing
> On the half rotten ash trees topmost twig,
> Beside whose trunk the gipsey makes his bed
> Up flies the bouncing wood cock from the brig
> Where a black quagmire quakes beneath the tread;
> The field fare chatters in the whistling thorn
> And for the awe round fields and closen rove,
> And coy bumbarrels twenty in a drove,
> Flit down the hedgerows in the frozen plain
> And hang on little twigs and start again.

Despite Clare's exhortations, heathland continued to be perceived by some poets as an emotionally sterile and empty environment. John Keats, for example, used the metaphor of the heath to reflect the bleakness of both his outlook and his soul in his 1815 poem *To Hope*:

> When by my solitary hearth I sit,
> And hateful thoughts enwrap my soul in gloom;
> When no fair dreams before my 'mind's eye' flit,
> And the bare heath of life presents no bloom;
> Sweet Hope, ethereal balm upon me shed,
> And wave thy silver pinions o'er my head.

Thomas Hardy's birthplace at Higher Bockhampton in Dorset. The vast heaths he knew as a boy, and about which he later wrote so evocatively, are now largely gone.

These conflicts notwithstanding, it was Thomas Hardy who was to become heathland's greatest literary champion. Born in 1840 on the very edge of the great Dorset heaths, he grew up with an intimate knowledge of the landscape, its ways and its people. Although he referred to the heath in many of his works, the fascination he felt for it is most potently crystallised in his novel *The Return of the Native*, first published as a serial in 1878. The book is driven by the power and immensity of Hardy's fictional 'Egdon Heath', against which the hapless characters play out their lives. The epic character of the heath is such that it takes on the form of a pagan deity, elemental and omnipotent, but not always visible: 'It could best be felt when it could not clearly be seen, …'. Furthermore, the heath has a defined personality, capable of passion but also of enduring blackness:

> The face of the heath by its mere complexion added half an hour to evening; it could in like manner retard the dawn, sadden noon, anticipate the frowning of storms scarcely generated, and intensify the opacity of a moonless midnight to a cause of shaking and dread … Only in summer days of highest feather did its mood touch the level of gaiety … Intensity was more usually reached … during winter darkness, tempests and mists. Then Egdon was aroused to reciprocity; for the storm was its lover and the wind its friend.

Moving across heathland was often difficult, owing to the rough tracks and generally primitive modes of transport. This photograph dates from the 1890s and was taken in Dorset.

If Egdon Heath is both the stage and the director, the people living on it are its puppets. For Hardy, the heath people were decidedly special, marked out by a deep understanding of their landscape and environment and by an empathy with nature and to the elements that the rest of humanity had lost. His characters were so bound to the heath that they were almost one and the same; of Clym Yeobright he says: 'He was permeated with its scenes, with its substance, with its odours … He might be said to be its product'. In many senses the heath is Clym's god, and indeed, the notion that heathland folk were somehow outside normal Christian boundaries still stuck in Hardy's time. Reference to their pagan, mystical powers and knowledge reappeared frequently through the centuries and has a defined presence in Hardy's novels. It was anticipated in the earlier works of George Borrow, whose *Lavengro* (1851) is a fascinating account of 'life on the road', and includes encounters with the gypsies who had made the heath their home. Certainly Borrow was the first writer to describe everyday gypsy life without lurching into bouts of excessive romanticism, and Hardy treats heathland folk in a similar fashion. Both these authors mark a significant turning point in cultural attitudes to country people, and for the first time, heath-dwellers were presented in Hardy's novels – mostly, at least – as intrinsically worthy and honest, even if they were caught up in a tragic spiral of events. Tellingly, this spiral usually

Gypsy caravans on the heath at Millbrook, near Nutley in Ashdown Forest, photographed in 1895. Traditionally many heaths supported gypsy communities, which were increasingly forced to resort to such areas as other options became closed to them.

has its root outside the heath, coming into play when the heath people have left their natural environment and spent time beyond its physical and spiritual boundaries (as with Clym Yeobright, for example), or equally when an outsider comes onto the heath.

Much of the power and value of the relationship between the heath-dwellers and their environment lay in a secret knowledge of the ways of nature, to which 'civilised' folk were sometimes forced to resort when their problems could not be solved by more orthodox methods. Hardy explored this idea more explicitly in *The Withered Arm* (1888), in which Gertrude Lodge, in desperation at the deteriorating condition of her afflicted limb and against her 'better' judgement, seeks out the help of the mysterious Conjuror Trendle, a dealer in furze, turf and other local products on Egdon Heath, but essentially a magician – 'he had powers other folks have not'; he also makes an appearance, at least in reference, in *Tess of the d'Urbervilles* (1891). The value placed by writers such as Hardy on traditional country wisdom – almost invariably unwritten and passed from father to son and mother to daughter – was soon realised in other fields. Foremost among these were traditional folksong and dance researchers, such as Francis Child and Cecil Sharp, who were active during the late nineteenth and early twentieth centuries. They focused particularly on gathering oral material from the gypsy communities, who with other travellers were commonplace on heathland until well into the last century. Some would probably still reside there now if they were only allowed to. Indeed, the New Forest had a substantial gypsy community until relatively recently. Needless to say, popular attitudes to such groups were generally negative and helped sustain and reinforce a lingering antipathy to the largely itinerant communities forced to live on the heath, mostly now resettled and divorced from their traditional landscape.

The late nineteenth century witnessed not only the stirrings in some quarters of an appreciation for heathland people, but also the genesis of a repoliticisation of heathland, following an earlier focus during the period of Parliamentary enclosure (see p.49). By now the age of Victorian philanthropy was well advanced, and with it the notion that open space should be retained and provided for the edification of the urban masses. This was naturally most important near the major cities and it was the carving up of the Surrey heaths – highly accessible to London since the coming of the railways – that sparked all sorts of protest. The Commons Preservation Society (now the Open Spaces Society) was at the forefront of the campaign, having already successfully rescued Hampstead Heath from being swamped by a sea of housing. Among its leading lights were Sir Robert Hunter and Octavia Hill who, with Canon Hardwick Rawnsley, were later to be instrumental in the establishment of the National Trust in 1895. One of the main focuses of their campaign was Hindhead Common and the Devil's Punch Bowl, which was later bought by the fledgling National Trust as one of its first major land acquisitions. Figures such as Hill and Hunter were closely associated with William Morris who, in 1884 in the radical political journal *Justice*, had railed against the surrender of the heathland landscape to agriculture, bricks and mortar:

> ... the grip of the land grabber is over us all; and commons and heaths of unmatched beauty and wildness have been enclosed for farmers or jerry-built upon by speculators in order to swell the ill-gotten revenues of some covetous aristocrat or greedy money-bag.

The activities of Hill and Hunter *et al* are undoubtedly a benchmark in heathland conservation and helped forge a new value for open heath as recreational space. However, with some slight modifications, Morris's words still rang true for a further century, and until very recently heathland was still being destroyed by a combination of unscrupulous owners, developers and unsympathetic (or corrupt) planning authorities. Indeed, the continual assault on sites such as Canford Heath in Dorset, once remote but now on the very edge of the burgeoning sprawl of Poole and Bournemouth, has only recently been arrested – and only after half of the heath has been lost to housing. Even now one does not have to look far to find negative views of heathland persisting in that part of the world; campaigners in favour of a new by-pass that would have carved through nearby Holton Heath were both shrill and flippant in their response to the proposal's rejection by a government inspector: 'People matter more than snails and the heathland!' came the cry. Yet local support is more often supportive than not, and is increasingly influential. Eleanor Cooke, in her excellent *Who Killed Prees Heath?* (1991), explores the fate of a bruised and battered Shropshire heath: loved and valued by many local people, but slowly dying from the neglect and disrespect of those who control its fate. Cooke proposes that its fate is symbolic of other breakdowns in society generally, and she may very well be right.

The world may have moved on in the last decade, and certainly heathland as a landscape type is more understood and protected now than at any other time in history. Yet much of this newly acquired status is derived from heathland's value to wildlife. Its contribution to cultural and social history is still imperfectly appreciated, and arguably a more immediate cause for concern. No one lives off the heath anymore, and those that remember doing so are ever fewer in number. First-hand understanding of what the heath provides and embodies will soon be a thing of the past. Instead, heaths are now looked after by wardens in Land-Rovers, and by conservation volunteers. Yet, in many senses they remain marginalised, places for dumping rubbish, for illicit sexual encounters, doing drug deals and starting fires. Against this backdrop it is no surprise, therefore, that the future of Prees Heath – and others like it – is still uncertain. Heathland may have more fans now than ever before, but it certainly needs them.

New housing encroaching onto Canford Heath near Poole. Despite the site's protected status, construction work continued here until very recently and has seriously compromised the integrity and condition of the heath and its wildlife.

Chapter Two

❖

Man on the heath

Heck cattle grazing on the Oostvaardersplassen in the Netherlands. Large herbivores similar to these would have been commonplace in Britain thousands of years ago, and would have helped maintain a patchwork quilt of forest, grassland and heath.

PREVIOUS PAGE Dealing with the challenging terrain called for ingenuity on the part of those who made the heath their home. On the heaths of Gascony in the south west of France, for example, shepherds took to using stilts. The extra metre of height allowed them to see further and thereby safeguard their flocks more effectively, and the stilts also allowed them to move with greater ease across boggy ground.

MAN ON THE HEATH

The history of Britain's heathland is one of human involvement. The same can be said of all British landscapes of course, but it is particularly so with heaths. Man has been present on the heathland scene for at least 10,000 years, and his activities there have helped define their evolution, distribution and characteristics. The vegetation on the heath may be natural, but the context in which it grows is decidedly shaped by the hand of man. From the earliest days of Mesolithic hunter-gatherers to contemporary countryside rangers, humans have been variously exploiting, destroying, managing, protecting and restoring the heathland environment and the wildlife it supports. Politics and economics have both been played out endlessly on the heathland stage, with the fate of this form of landscape subject to the vagaries of agricultural prices and trends, and to the demands of political expediency and changing lifestyle patterns. In more recent times heathland has become the focal point for renewed political energies, the battleground for conservationists and developers alike. It is one of our most familiar yet contentious landscapes.

The beginnings – forest or savannah?

The precise origins of heathland have long been a source of speculation. For many years the assumption was that virtually all heathland was once forest. This was based on the premise that following the last Ice Age, which ended in c.11,000 BC and during which the British Isles were covered by tundra, improving climatic conditions across much of Britain soon led to the land becoming covered almost entirely by primeval forest. This so-called 'wildwood' was comprised mostly of oak, lime and hazel, and was presumed to have extended from coast to coast in a seamless robe of impenetrable green. In this scenario, only locations too high, too dry or too exposed to support the growth of trees were free of forest cover. These limited areas were mostly taken up by natural grassland and were deduced to constitute the early home of plant species that were later to colonise areas of open country, including heathland.

The huge tracts of mixed deciduous forest remained largely unaffected by humans until Mesolithic times, some 7,000 years ago. The Mesolithic hunter-gatherers began to make temporary clearings in the wildwood, possibly to encourage the development of better grazing, which would be attractive to the wild animals they hunted. We know from the archaeological record that large numbers of herbivores roamed Britain at this time, including aurochs (wild ox), elk, red deer and tarpan (wild horse). Where these animals became concentrated in large numbers, as in forest glades, they would prevent tree regeneration by browsing and grazing off the new growth. Open areas of grassland were therefore maintained, and some of these may have developed heathland characteristics. Even though the Mesolithic human population in Britain was fairly small, and its impact on forest cover likely to be correspondingly modest, the roots of 'manmade' heathland almost certainly lie here.

However, it was not until around the fifth millennium BC, when the more numerous early settlers of the Neolithic period began to establish more permanent settlements, till the soil and pursue animal husbandry, that the human onslaught on the forest began in earnest. Landscape historian Oliver Rackham identifies this as the time when 'they set about converting Britain and Ireland to an imitation of the dry open steppes of the Near East in which agriculture had begun'. Trees were steadily cleared off the land to make way for pasture for grazing animals and fields for the growing of crops such as emmer and einkhorn wheats, barley and plantains. The rate of clearance was faster in some places than in others; in the densely populated Brecks of East Anglia most forest had gone by the end of Neolithic times, whereas areas such as south Dorset remained quite heavily wooded until well into the ensuing Bronze Age.

Stripping the land of its forests was no mean undertaking, with only primitive tools such as stone axes to hand, and potentially, the aid of fire. However, the latter could never have been used so effectively in the British climate as it was by the early peoples of North America to clear their forests, for example. Sheep, goats and cows would

Old flint pits at Grimes Graves in the Norfolk Brecks. This area was cleared of its trees during the Neolithic period and then mined for flints. The resulting landscape is typical of the stony, grassy heaths of Breckland, which although unsuitable for sustained arable cultivation, were able to support large flocks of sheep.

The original landscape of much of Britain was probably more one of wood-pasture than of continuous forest. Closed canopy woodland would have been interspersed with areas of scrub, heath and grassland, as here at the National Trust's Hatfield Forest in Essex.

have helped prevent the regrowth of trees and shrubs, but the destruction of such extensive areas of forest remains an outstanding achievement by so few with so little. Overall, Rackham estimates that half of the wildwood had gone by the early Iron Age, c.500 BC. Certainly by Roman times, five hundred years or so later, there was scarcely much more forest in England than there is now.

Initially the cleared areas would have been taken into rudimentary arable production. However, the clearance of trees exposes the ground below to leaching by rain, and this process is especially severe in poor soils, where any nutrients are soon washed downwards, out of reach of crops. When this happened, and crop yields fell accordingly, cultivation would have been abandoned in favour of grazing, with burning regularly employed as a way to stimulate new growth. The manure of the animals would have helped temporarily re-fertilise the denuded soil, but the process of leaching and impoverishment would continue until the land was so acid that it could only support plants tolerant of such conditions, notably heathers and certain grasses. The emergence of such vegetation marked the advent of heathland. However, even after such deterioration, the land continued to provide limited grazing for livestock, which both prevented the regeneration of forest and helped ensure that heathland remained an integral part of the agricultural scene.

Soil analysis shows a marked increase in heather and grass pollen during Neolithic times, when land was being extensively cleared and turned over to grassland and, eventually, to heath. That many heaths were already in existence by the late Neolithic and Bronze Age periods is made clear from the pollen evidence taken from beneath contemporary burial chambers or barrows: it is often high in classic heathland indicators, including heather, and in grassland species such as bird's-foot trefoil and sorrel. Such plants were presumed to have colonised these open areas from the natural grass- and heathland that existed on those exposed sites that had never been cloaked in woodland.

However, during the late twentieth century, authorities increasingly questioned the presumption that the whole of Britain had once been a virtually continuous forest. First, how could populations of large herbivores have existed under the closed canopy? The vegetation would not adequately have met their grazing and browsing requirements, and they could not have thrived solely in the limited – and unpredictable – glades created either by man or by natural events such as lightning strikes, fires and gales. Equally, is it correct that these large mammals simply had no effect on the forest, which remained overwhelmingly closed until man's advance? Thirdly, just how easy would it have been for heathland flora to colonise the cleared areas of forest so comprehensively from its few and distant 'reservoirs' elsewhere? Fourthly, and most intriguingly, how does the concept of the continuous closed forest square with the success in Britain of the oak, which requires open conditions in which to achieve successful maturity? The spread of both hazel and oak trees is dependent on the behaviour of birds such as the jay, which carries nuts and acorns away from the parent trees and buries them in turf in open areas. How could this behaviour have evolved if there were no open spaces in which to do this? The conclusion is increasingly that the early landscape of Britain was not so much one

great forest but more probably a vast and dynamic landscape of wood-pasture: areas of open grassland interspersed with scrub, groves, scattered trees and patches of thicker forest. Landscape ecologist Frans Vera proposes that 'the original vegetation in the lowlands of Europe is a park-like landscape where the succession of species of tree is determined by large herbivorous mammals and birds such as the jay, which act as facilitators for certain species of tree'. From the earliest times in this European 'savannah', glades would have opened up as a result of disease, pests, storms and tree-felling by beavers; these clearings would have been maintained and enlarged by the grazing and browsing of the large herbivores, who therefore play a central role in the shaping of the landscape. On the grasslands that result from this intensive grazing, patches of thorny scrub develop, unpalatable to the herbivores and so serving as a protective nursery for young trees. These trees develop within the scrub, eventually forming a grove, over which the tree canopy eventually closes. The scrub retreats in these shaded conditions, allowing renewed access to the herbivores, which then graze the ground below the canopy once more, preventing any further regeneration. The trees mature, die off, and a glade opens once more within the grove, eventually becoming grassland again. So the cycle continues. At any one time, therefore, this early landscape would have been composed of a series of groves, glades and grasslands, each at its own particular stage in the cycle, with scattered stands of thicker forest in those places inaccessible to large herbivores and therefore 'out of cycle'.

Within this mosaic heathland is likely to have been a regular – and fully natural – component of the landscape, especially on poor soils that were heavily grazed. When the appropriate opportunity arose, with the opening up of glades within woodland, heathland plants would colonise readily from adjacent areas as, indeed, they do today when conditions are right. The development of tracts of heathland would therefore have been part of the development of grassland in the landscape cycle, and the dormant seed bank in the soil would readily revive from the last 'heath/grassland' stage in that cycle. If this theory stands – and the evidence is persuasive – then the wildwood was not a wood at all and heathland cannot be attributed as a wholly manmade landscape. It was always there, and all man has done is shape it, use it and replicate it.

Heaths under the Anglo-Saxons and the Normans

By Anglo-Saxon times (roughly from the late fifth century AD to the Norman Conquest of 1066) heathland was very much part of the working landscape. This is clear from its integration into contemporary agricultural and social systems – references to heaths appear in charters and tithe deeds, for example – and from the incidence of Saxon place-names based on heaths and on the plants found upon them. Rackham identifies no fewer than 29 English settlement names containing variations of the word 'heath', including Hatfield, Hadley and Hatton ('heath-town'). A further hundred place-names suggest links to heathland via typical plants, such as gorse, broom – Bromley, for example – and bracken or the Old English *fearn*, which appears in names like Farnborough.

Land was the critical factor in Anglo-Saxon relationships. The fate of all sectors of society was directly dependent on the fortunes of agriculture and, thereby, on control over the

Draught animals such as oxen were an essential component of early agriculture, and the provision of adequate pasture for them was a high priority. When not being used for activities such as ploughing, the beasts would be turned out to graze on areas of manorial waste, much of which was heathland.

soil. Life was centred around the manor house, the owner of which held 'lordship of the land' and exercised control over the surrounding territory, both land within cultivation and land without, subject to a complex but generally respected apparatus of reciprocal rights that involved all members of the local community.

Central to the contemporary system of land management was the concept of 'waste'. Land was divided into several categories, depending on use: arable, pasture, meadow, forest and waste. The latter did not have the connotations we attach to it today; far from being redundant or useless, it was simply land that was 'extra', usually unenclosed, and treated as a source of supplementary grazing and of other useful products such as fuel. Every parish would contain areas of waste, often comprising a range of different landscape types over which there was common access. Heathland was a prominent component of waste, and was often associated with areas of wood-pasture – further reinforcing the argument that the original landscape of Britain was one of quasi-parkland (see above). Furthermore, heathland waste was particularly valued for the range of products it could offer, and especially for its free grazing. The heaths would have been regularly grazed by large numbers of livestock, including cattle, horses, donkeys, sheep, goats, pigs and even chickens and geese.

However, towards the end of the Saxon period an increasing number of such 'commons' – many of which were heaths – were sequestered by the aristocracy and incorporated within their manorial holdings. The Norman barons continued this trend, although the Domesday Survey of 1086 contains only one explicit reference to heathland ('*bruaria*'). This is probably explained by the Survey's emphasis on formal assets, rather than the more nebulous value of waste. Meanwhile, local people continued to be allowed certain rights over land purloined by the nobles, even if any concept of common ownership was effectively extinguished. Such property therefore continued to be regarded as common land, and despite some *ad hoc* exclusions during the sixteenth century in particular, the overwhelming majority of heaths remained common land until the age of Parliamentary enclosure during the late eighteenth century and first decades of the nineteenth. Many still have common rights over them today.

Common rights

The concept of 'rights of common' goes back to at least Anglo-Saxon times. Essentially a system of land management based on the granting and exercise of specific rights to use the land in particular ways, it has played a central role in the history of heathland. Whilst heaths are generally poor in terms of soil, the holding of common rights over them has traditionally provided a valuable source of income to the right-holders, many of whom held only modest smallholdings and so relied heavily on their rights elsewhere. These rights were administered by the manorial courts, the objective ostensibly being to ensure an equable and sustainable apportionment of local resources. However, the process was, not surprisingly, driven by the interests of the manor; for example, the right to graze oxen on the manorial waste was granted in return for the oxen ploughing the manorial fields for a certain number of days per annum. Originally, rights were usually attached to properties rather than to individuals. This situation was to change in later years, when the sale of common rights became a regular occurrence in some localities.

On heathland the most significant rights were those of pasture, the right to graze animals (cattle, sheep, horses and goats); of turbary, the right to dig turf (and, in some places, peat) for fuel and/or roofing; of estovers, the right to collect wood for fuel and fencing, usually off the ground, but also including gorse underwood and wood cut from dead and decaying trees; and finally, the right of common in the soil, *viz.* the extraction of gravel, sand, stone and other minerals. Some, more wooded, heaths also offered rights of pannage, whereby right-holders could run pigs in autumn to feed on acorns and beech-mast. In most instances right-holders were restricted or 'stinted' to a certain number of livestock and/or to a specific period of time. Generally speaking, an individual could hold a range of different rights, although many were highly specific and not readily transferable; for example, the right of estovers could be restricted to mean the taking of only sufficient fuelwood for use on the original hearth of a dwelling; if the hearth was moved, then the right lapsed. Equally, refusal or inability to exercise rights could make them vulnerable to abrogation.

The definition of rights varied quite widely from place to place; in the New Forest, for example, turbary rights were sometimes interpreted to include the 'cutting of fern, heath and furze', although the latter was more usually regarded as a customary practice: i.e. not expressly formalised as common rights were, but still considered a right in all but the legal sense. Such practices were as vital to heath-dwellers as common rights, and in many places took on semi-formal status, even down to individuals or families having an annual 'claim' on cutting a specific tract of gorse or bracken. In addition, there was also extensive illegal exploitation of heathland by those who held neither common rights nor any claim on customary practice.

Sheep were often grazed on heathland during the day and penned at night to prevent them wandering. Their dung was a useful source of fertiliser for the generally poor soils.

Heaths from the Middle Ages to *c.*1750

The Middle Ages saw the break-up of the manorial system and the beginnings of agriculture as a profit-making business. This had implications for heathland, as it brought with it the notion that land could now be a commodity for sale, a source of income and not simply one of power. 'Waste' could be turned into cash if it could be made into useful agricultural land. During the thirteenth century the process of sequestration that began under the Saxons gathered pace, with many landowners consolidating and enclosing their demesne lands, and focusing particularly on untilled waste – much of which was heathland. Such areas of land, known as assarts, were often then cultivated and/or let out to the new and growing class of tenants, many of whom were former bondmen, now freed from feudal ties. Great landowners such as the 1st Lord Berkeley, who held his family estates in Gloucestershire from 1281–1321, actively promoted farming progress and the enclosure of waste, and many of his peers encouraged their tenants to annul existing rights of common over cultivated land (it was regular practice to graze livestock on arable land once the crops had been harvested, for example) in favour of a system of adjustment – i.e. exchanging money or property in return for the revocation of rights – and the establishment of separate farms.

The gradual erosion of feudalism also spawned a whole host of social changes that were to have implications for the ways in which heathland was both used and regarded. The exchange of tied labour services for their cash value may have been the genesis of an enterprise society but, as agricultural historian Lord Ernle pointed out in his analysis of change during the fourteenth century in *English Farming, Past and Present* (1912), it was a time in which 'the struggle for life becomes intensified; the strong go to the front, the weak to the wall; for one man who rises in the social scale, five sink'. Certainly, many men were able to move outside the collapsing manorial system into a more independent life based on the proliferation of smaller holdings that were forged from the disintegrating demesnes. But others – mainly labourers, as opposed to the increasingly successful tenant farmers – found themselves losing out, no longer part of a reciprocal system that, whilst not perfect, did at least provide a degree of security and communality that was now clearly on the wane.

The result of these changes was an increase in the number of people relying on those common rights that did survive; many of these were over heathland, land that was traditionally difficult to bring into cultivation, despite the best efforts of the early improvers. As enclosure grew apace, so heathland became increasingly a refuge for the needy and dispossessed. It is likely that this period saw an increase in the number of people actually living on heaths, as in many cases they had lost their tenure on the more productive land. Such landless labourers relied on selling their skills and time to whoever would buy them. Meanwhile, attempts continued to bring heathland within the agricultural envelope. For example, on the sandy heaths of inland Norfolk and Suffolk, a type of shifting cultivation was practised, whereby areas of heath – known as brecks, from which the area has taken its name – were periodically 'broken', ploughed and planted up with crops (usually root vegetables or barley). Once the few nutrients in the soil were exhausted, which generally took only two or three years, the land was abandoned and allowed to revert to heath. Meanwhile sheep, which would roam the heaths by day, were corralled at night onto fallow land or land proposed for imminent cultivation, their manure providing valuable enrichment. In the Middle Ages many thousands of sheep were maintained in Breckland in this way, often in huge flocks, and most of them owned by the local abbeys. The native breed – the Norfolk Horn – was well suited for life on the heath, being wild by nature and long-legged, which helped it move across the sandy wastes. An anonymous old rhyme alludes to its tough character:

> Old Norfolks will serve well enough
> To dung our sand – unless until
> They prove that some outlandish breed
> Are hardier and can further scramble
> With strong bare legs and bellies high
> Through brakes, broom, furze, heath wet and dry.

Sheep-grazing on heathland may have been common practice across Britain in the Middle Ages, particularly with the growth in the wool and textile trade, but cultivation was not; the Brecks were unusual in that sense. Farming activity was usually restricted to the more fertile areas either adjacent to, or bisecting, heathland – on which the soils were considered too unforgiving to merit the effort of ploughing and crop husbandry. The answer, of course, lay in 'improving' the soil and whilst the presence of grazing animals helped, it was a decidedly short-term measure with no lasting impact.

Although large-scale soil improvement did not become common until much later, some pioneering landowners encouraged the few techniques that were understood at that time. Foremost among these was marling, which had been practised as early as the Roman period and was applied regularly, if patchily, across many parts of England, especially on marginal land such as heathland. Marl is a calcareous clay, rich in nutrients and which is particularly effective at both helping consolidate loose, sandy soils and at neutralising the high acidity typical of heaths. Seams of marl are often present on heaths where chalk overlies acid soils. It was dug out in lumps from pits excavated for this purpose, carted onto the fields and spread evenly across the ground. Depending on the amount used, it could enrich a field for up to a hundred years and so was greatly valued, to the extent that it was sometimes specified under the right of common in the soil (see above). In some regions it also had another use: as a building material in cob construction (see p.79).

One particular event stands out during the medieval period as having particular – if only temporary – significance in the history of British heathland: the Black Death of 1348–49. The crash in population, down by between one-third and one-half in many areas, led to a severe reduction in agricultural activity. Fewer farmers meant less livestock and a rapid decline in grazing; this would have led to many heaths becoming overgrown by scrub and trees, and it is likely that the overall extent of heathland in Britain was reduced substantially as a result. Records indicate contemporary increases in tree pollen and decreases in pollen from grasses and mosses, evidence that the open areas favoured by the latter were disappearing as the scrub advanced. It most probably took several decades for the agricultural population, grazing levels and heathland distribution to be back to something like their 1347 levels, although it is worth noting that rural depopulation did give larger landowners the opportunity to reclaim areas of waste and use them more intensively for their own purposes.

During the Middle Ages and well into the Elizabethan era, much of Britain's heathland was found within the confines of what were nominally 'forests'. Many of the great forests of England, including Sherwood, and most notably the New Forest, contained extensive tracts of open heath with very few trees, if any. Many of these forests were originally established as hunting preserves for the monarch and his or her nobles, but the resources within their boundaries were utilised in much the same way as those outside. For example, local residents would hold rights of estovers and of grazing within the forest, and in this way the forest heaths were maintained and, if grazed heavily, even extended. However, as the system of royal forests began to break down during the seventeenth century, and encroachment by the burgeoning gentry class grew, so yet more heathland – and the useful commodities it provided – was taken into private ownership and increasingly denied to those less wealthy and influential.

The pattern throughout this period is one of heathland's continued position as a valuable commodity, 'worth not much less than arable', as Rackham rightly says, but one that was subject to the vagaries of changes in land use and evolving patterns of ownership. The sixteenth century in particular was a time of widespread conversion of

Thomas Coke, later the Earl of Leicester, was one of Britain's foremost agricultural improvers, his estate at Holkham in Norfolk serving as a showpiece for new techniques and livestock breeds. Yet the advances he pioneered spelled disaster for many heaths, swept away in the face of more effective farming methods and a more regimented approach to land ownership and management.

tillage to pasture in many parts of England, and the increase in overall numbers of grazing beasts would both have helped perpetuate existing heathlands and create new ones in wooded areas that would not have been grazed heavily (or at all) in earlier periods. Meanwhile, heathland remained overwhelmingly a landscape of shared use – along with downland and moorland, for example – albeit one that in terms of ownership was increasingly gathered into the hands of the few.

The Impact of Agricultural Improvement and Parliamentary Enclosure

The late eighteenth century was a period of dramatic change in Britain. Ambitious programmes of land improvement brought large areas of England and Wales into agriculture for the first time. The introduction of new, stronger strains of crops, more refined farming techniques and better land husbandry, as well as early forms of mechanisation, totally revolutionised how land was farmed and, perhaps more tellingly, the ways in which it was perceived by those who owned it. Meanwhile, the advent of industrialisation prompted both a drift away from the land towards the burgeoning cities and towns and a general increase in population, which itself created new demands and tensions.

These developments had a serious impact on heathland. Agricultural improvement made it possible to bring into profitable production land that would otherwise have been given up on as worthless, or at the very most farmed on a desultory basis. Naturally enough, this was usually marginal land, and much of it was heath. 'Troublesome' land such as this was a particular target for zealotry, the degree of 'waste' in Britain being regarded by some as nothing short of a national disgrace. In his *Observations on the Present State of Waste Lands of Great Britain* (1773), agricultural commentator Arthur Young was clearly affronted by the amount of uncultivated land that covered much of the country, and two years later Nathaniel Kent, one of the leading 'improvers', was horrified to find 'that within thirty miles [of London] there is not less than 200,000 acres [81,000 hectares] of waste land' – most of which was heathland.

The attack on heathland waste was launched on philosophical as well as practical grounds. The fact that many heaths were also commons prompted additional – and special – venom from some quarters, with one early eighteenth-century commentator branding commons the 'seminaries of a lazy Thieving sort of People'. Regarded as the home of the idle and dissolute, their destruction was therefore considered by some as a highly desirable form of progress. At the same time, communications were improving, making travel easier. Even the more remote heaths were not so isolated any more, their inhabitants increasingly vulnerable to interference from outside.

The opportunities provided by new crops, improved methods of cultivation, advances in soil science and more sophisticated stockbreeding changed the face of Britain. The traditional pattern of land ownership had been for a network of smallholdings, each made up of often widely scattered piecemeal property, and the owners consuming most of what they grew. Much farming took place in open fields made up of strips, each owned by a different individual, who was also entitled to grazing rights over nearby common land. Such a system was clearly complex and, to the new way of thinking, inefficient and wasteful. Farms were increasingly consolidated in the quest to maximise the potential of the land, with the larger landowners actively acquiring smaller parcels of property from their neighbours. The late eighteenth century consequently saw the advent of the first 'superfarms', large concerns where agriculture was a commercial enterprise rather than simply a means to support one's immediate family or community. Large estates such as Raynham Hall in Norfolk (where Lord Townshend pioneered many advances in crop rotation and cropping systems) came to dominate the landscape, both physically and psychologically. It is estimated that between 1696 and 1793 some two million acres (over 800,000 hectares) of unused land were added to Britain's agricultural estate. Most of this was turned to arable production, with the formerly lucrative wool trade having declined during this period and cereal prices now performing more strongly. Much of the demand for the latter was fuelled by the growing populations of the industrialising cities.

The desire to bring land into production – labelled by one contemporary commentator as '*Terramania*' – spawned some powerful acquisitive forces. Thirsty for land, the expanding gentry class saw it as their duty to 'enclose and improve', and the concept of enclosure by consent was increasingly considered too time-consuming and problematic. Dramatic alternatives were proposed, namely enclosure by Act of Parliament. Such acts facilitated the enclosure of land in particular parishes, with the drawing of boundaries and division of ownership largely determined by those holding most of the land: a system that allowed large landowners to steamroller the concerns and claims of their smaller neighbours. The impact of this carve-up was dramatic: as landscape historian Tom Williamson has pointed out in *Hedges and Walls* (2001), '... by 1850 almost all open fields in England and Wales, and the majority of common land, had disappeared from the landscape'. Even the value of heathland for grazing was affected, for with agricultural improvements it was often more efficient to clear a heath, enrich the soil and plant it up with fodder crops on which the livestock could be fed.

However, heathland was not just enclosed and turned into agricultural land. Harking back to traditional fears about its wild, uncivilised quality, it also suffered from attempts to 'tame' it, its subjugation being actively pursued by certain landowners as an act of public benefit and philanthropy. One of the most outstanding examples of such largesse is the Dunston Pillar, an extraordinary monument constructed in 1751 by Sir Francis Dashwood (who was later to become Chancellor of the Exchequer) on the heaths just south of Lincoln. Originally 92 feet (28 metres) tall and topped by a lantern an additional 15 feet (4.5 metres) high, the pillar was erected to serve as a marker by which travellers could guide themselves across the featureless landscape of the area. At that time extensive tracts of sandy heathland ran on both sides of the Lincoln–Sleaford road (now the A15) and there had been cases of travellers losing their way and even of stagecoaches and carriages careering around on the heaths for hours, having lost their sense of direction. At night the brazier in the lantern was lit and the pillar became a bizarre inland lighthouse. At its base were a bowling green and an assembly room, the latter demolished by the mid-nineteenth century but once much used for card parties

The now beaconless Dunston Pillar was once a valuable guiding light to those travelling across the sandy wastes south of Lincoln. The heaths over which it once towered are long vanished, and it is now in a rather sorry condition, sandwiched between houses next to a busy road.

on race days at nearby Lincoln racecourse. During the winter of 1808–9 the pillar's lantern was damaged by wind and replaced by a statue of King George III, later removed. Meanwhile, the remains of the Dunston Pillar still stand, even though the former heaths it once dominated have all been destroyed and taken into agriculture. Interestingly, the area is still called 'the Heath'.

Elsewhere, heathland was destroyed to make way for cosmetic landscape improvements. Felbrigg Hall in north Norfolk, now in the ownership of the National Trust, was originally bounded to the north by extensive tracts of heathland. These were largely enclosed, emparked and planted up with shelter belts, which now form the Great Wood. One of the guiding influences in this landscaping during the 1770s and '80s was Nathaniel Kent, and the success of his conversion of wilderness to Arcadia was said to have influenced Humphry Repton in his own designs for nearby Sheringham Park, which he described as his 'most favourite work'.

The agrarian revolution and the success of Parliamentary enclosure sounded the death knell for many British heaths. Vast heathland tracts were fragmented, ploughed up or simply removed in the re-landscaping of a large estate. For example, the extent of heathland present in Dorset in 1760 (and shown in the maps of Isaac Taylor) had probably changed little in 2,000 years, yet by the time of the first Ordnance Survey in 1811–17 a quarter of this had already gone. The apparent luxury of leaving so much land seemingly idle was clearly more than the forces of progress could bear.

Yet heathland was providing a livelihood for many thousands of people, some of whom were pushed to the brink of destitution – or more likely, to the slums and mills of the cities – by the actions of the improvers and enclosers. There was occasional respite, however; for example, at the time of the 1805 enclosure award at Ringwood in Hampshire the Commissioners agreed to set aside over 400 acres (162 hectares) of heath as 'turf grounds' to compensate those who had lost turbary rights. Equally, some philanthropists realised the dire straits into which the landless had been forced by enclosure and took action accordingly. In the late 1840s Miss Georgina Talbot became concerned about the fate of poor families near Bournemouth, who were involved in smuggling and poaching on the heaths near the town. She purchased an area of heathland specifically for them and commissioned a local builder to lay out a series of cottages – rather deluxe by the standards of the day – which were offered to families at low rent, complete with grazing rights on the heath and the opportunity to grow their own crops. She later funded the construction of almshouses for those unable to provide

The Talbot village almshouses, which were endowed in 1862. In return for strict rules concerning their conduct, the inmates were allowed 6/- per week, as well as two tons of coal annually and the free attendance of a doctor.

for themselves, as well as a school and a church, which was not completed before she died. She became the first person to be interred in the new churchyard. The 'Talbot Village' was a form of social experiment, an attempt at creating a self-supporting village community. Many of the buildings still remain, although most of the heathland was subsequently sold off for housing or for educational purposes: Bournemouth University now occupies part of the site.

However, such acts of benevolence were hardly the norm. The advocates of progress were generally unimpressed by the objections of the dispossessed, citing the greater national good as the overriding concern. As Sarah Zaluckyj points out in *Hartlebury Common: A Social and Natural History* (1986), 'There is little room in this attitude for the importance of these areas to the local inhabitants, especially common right holders. The sense of personal loss as well as economic loss should not be forgotten ...' Many people lost not only their hold on the land at this time – unable to legally establish rights that their families may have held informally for many decades – but also any sense of self-worth and stake in society. Transformed into landless labourers, in desperation they became squatters, only to be forcibly evicted. Heathland was to become their final refuge.

1850 to the 1980s

The mid-nineteenth century saw the heyday of 'high farming', with sustained demand and high prices. The remorseless march of agricultural progress went on, mirrored by the continued conversion of areas of heathland into increasingly intensive agricultural use. For centuries Britain's heaths had been valued as a supplement to agricultural and rural life; now they were marginalised and derided, and seen primarily as targets for destruction and replacement. This began to have a serious impact on certain species of flora and fauna. As Tom Williamson remarks, 'The destruction of thousands upon thousands of hectares of ancient, semi-natural habitats – acid heath, chalk heath, downland and fen – was an ecological disaster on an awesome scale'. Open-country species such as the stone curlew came under real pressure, as their habitat became fragmented and increasingly regimented by fences, hedges and the paraphernalia of modern farming. Populations of birds and butterflies became isolated and thereby more vulnerable to collectors and further destruction of their habitat.

Temporary respite came during the agricultural recession that started in the late 1870s and from which most areas only really recovered with the advent of the Second World War in 1939. As far as heathland was concerned, the economic downturn in agriculture was a mixed blessing. Whilst lower land and food prices meant that there were fewer incentives and resources available for the reclamation of heathland, the reduction in farming activity – and of grazing in particular – did lead to many heaths falling out of regular grazing and thereby becoming overgrown by scrub. Throughout this period heathland continued to lose its place in the local economy. It is, of course, impossible to measure in cash terms the value of the benefits derived from common heaths by the rural underclass, or the loss inflicted on these people by the damage and destruction of heathland. The onslaught was not just from the physical removal of their economic landscape, a gradual process of attrition that had been taking place for some time; it was also one of general economic development, whereby the traditional handicrafts used to supplement the income of heathland dwellers were being increasingly superseded by a flood of manufactured goods from the cities.

Other pressures were also building. The open spaces of heathland had long been used for the assembly of crowds and particularly militia (Blackheath in London was a favoured setting for military reviews during the eighteenth century), but the mid-nineteenth century saw increasing demands for the establishment of a standing army. The War Department subsequently moved to create the conditions under which such an army could be accommodated and trained, and acquired extensive areas of heathland for military use: their remoteness, open aspect and low population density made them ideal. The main areas of acquisition were around Aldershot in Hampshire and in neighbouring parts of Surrey, where large areas of heathland are still owned by the Ministry of Defence. The heaths of Chobham Common were the site of Queen Victoria's first review of her troops, in 1853. Meanwhile, the empty and remote quality of the heathland landscape made it increasingly attractive for the testing and production of potentially dangerous substances; for example, at Eyeworth Lodge, near Fritham in the New Forest, a gunpowder factory was established on the open heath in the 1860s, using the extensive supplies of charcoal from the local woods. The factory lasted sixty years or so, and in its heyday had some seventy buildings. Hardly a trace remains today.

By the early years of the twentieth century the race for military preparedness and the need to trial new weapons and methods of warfare had led to further military acquisitions of heathland, a process that gathered speed even more dramatically with the outbreak of war in 1914. For example, the heathland around Elveden in the Anglian Brecks was used by the Machine Gun Corps – later the Tank Corps – for training, before they were moved to Bovington in Dorset. The Dorset heaths were subsequently to become a centre for military activity, with a full-calibre gunnery range established east of Lulworth Cove. When war ended in 1918, this particular area was compulsorily purchased by the War Office and remained in army use, although this was not necessarily the case elsewhere. Many sites were returned to their former use and owners, such as Prees Heath in Shropshire, home to an army training camp during the war but which was carefully returned to its original appearance by the army upon leaving in 1918.

However, the start of the Second World War spawned a new round of compulsory purchases from which many heaths never recovered. This was particularly so in East Anglia, where following the evacuation from Dunkirk in 1940 almost 48,000 hectares (118,500 acres) of poor-quality agricultural land, much of it heath, were commandeered and used to train the army for the eventual invasion of Europe. The flat, open terrain of the region's thinly populated heaths was equally ideal for the establishment of airfields. Some of these – such as Lakenheath in Suffolk – remain in military hands, whilst others – most notably East Wretham Heath, now owned by the Norfolk Wildlife Trust – have been relinquished by the Ministry of Defence and restored to something like their pre-war condition. Military ownership of such places continues to be contentious, most famously at Tyneham in

Dorset, where the army has still not left, despite promises made to the owners upon occupation in 1943. Paradoxically, military ownership and control over extensive tracts of heathland may have been the landscape's salvation, as it has at least ensured its protection from the sort of overbearing pressures that have devastated heathland elsewhere; as a result, many military-owned heaths are notable havens for endangered wildlife (see p.132). A range of intense civil pressures also came to the fore during the twentieth century, in the form of growing demand for land for housing, industry and infrastructural projects such as road improvements. Housing pressure began in some areas as early as the late nineteenth century, when improved rail links made parts of Surrey more directly accessible to commuters from London. Elegant 'gentlemen's homes' began to spring up on the fringes of heathland in Surrey, for example, and many formerly remote rural areas soon became part of an expanding commuter belt. In many respects heathland was ideal for house-building: generally flat or easily landscaped terrain, with plentiful sand and gravel on hand. Land acquisition costs were usually modest, and so it is no great surprise that throughout the twentieth century extensive areas of heathland were

Military camp at Snarehill near Thetford in Norfolk, September 1912. The camp was set up on an area of sandy heathland, and the squadron of planes belonging to the newly formed Royal Flying Corps attracted hundreds of spectators.

A 1920s view of a Breckland heath, ploughed into strips prior to the planting of conifers by the Forestry Commission.

transformed into housing estates. This was a process that was really only curtailed towards the end of the 1980s, when developers were still being granted permission by the local planning authorities to build housing on Canford Heath, near Poole in Dorset, despite its well-documented wildlife value and notification as a Site of Special Scientific Interest. With housing come roads, and the construction of new road networks – together with the widening of existing highways – has been a major source of heathland destruction and fragmentation in recent decades. Nowhere is this more readily apparent than around the Poole–Bournemouth conurbation, where an endless sphaghetti of by-passes, relief routes and feeder roads has eroded the remnants of those heaths already smothered by concrete and cement.

The twentieth century also saw the transformation of heathland from being a working landscape to one of primarily recreational use. This was especially the case on the heaths of London, which, as the city spread, became increasingly marooned in an ocean of housing. Their farms engulfed and destroyed, the sheep and cattle on areas such as Blackheath and Hampstead Heath were replaced by increasing numbers of people, drawn there not for work but for pleasure. In many instances this led to the establishment of formalised sporting facilities, most notably golf courses. The growth in popularity of golf during the early twentieth century led to a proliferation of courses and links across Britain. Again, heathland's open aspect and light sandy soils made it ideal for this sort of use and, whilst much has been made of the landscape component of the golfing environment, the destruction of most of the original heathland habitat did incalculable damage at most such sites.

Twentieth-century heathland also suffered at the hands of Britain's crisis over timber. Heaths had always been vulnerable to tree-planting, and Rackham has highlighted the impact on the Surrey heaths of the self-sown descendants of Scots pines planted in the nineteenth century. However, the First World War revealed a severe home shortage of decent-quality timber, and as a result the Forestry Commission was established in 1919 to address the problem. Charged with implementing a proposal to plant up some 72,000 hectares (178,000 acres) with trees over the following eight decades, the Commission soon set about its task. Over the next few years, large blocks of Breckland were purchased and prepared for planting. Some of the land was in a semi-derelict state, in agricultural terms at least, but much was prime heath, rich in wildlife. In the late 1920s and 1930s the hard work involved in preparing the ground and planting the seedling trees was carried out by gangs of unemployed men, kept in 'Transfer Instructional Centres', highly regimented units built in isolated locations out on the heaths and essentially little better than labour camps. The efforts of their inmates helped transform one of England's most notable open heathland landscapes into one of brooding coniferous plantations. The Forestry Commission also planted heavily on the Sandlings of Suffolk and on the heaths around Poole Harbour in Dorset which, together with Breckland, were the three most extensive remaining heathland areas in England in 1920. Some 45 per cent or so of their total extent was subsequently lost under trees in the ensuing decades.

By the 1980s British heathland was in crisis. Hammered by several decades of neglect, abuse and outright destruction, it was reduced to only 40 per cent of its 1945 extent, having been built over, ploughed into arable production, carved by roads, smothered in fir, pine and spruce, eaten up by out-of-town supermarkets and industrial estates, churned up by mountain bikers and four-wheel drive enthusiasts, and belittled by greedy politicians and developers. A victim of social and economic changes, the heathland landscape had become redundant and disconnected from everyday life.

Chapter Three

❖

Living off the heath

LIVING OFF THE HEATH

For centuries heathland had provided man with a living. People would either earn their living directly off the heath, or would gather raw materials there and process them into products for sale elsewhere. Heathland offered a huge variety of opportunities to provide for one's own needs and to make a little money, too. However, most heathland dwellers (often known as 'heathcroppers') lived at barely above subsistence level and relied on the little left after their own needs had been met to purchase goods that were not available on the heath. Life was hard, and much effort was required to make ends meet; heaths could therefore be surprisingly busy places, with local people carrying out all sorts of tasks there throughout the year.

However, by the mid-twentieth century, changing socio-economic patterns and higher standards of living meant that this direct reliance on nature and the landscape had all but ceased. The uses and products that had made heathland such a valuable asset for thousands of years were no longer required; as demand fell, so the people whose livelihoods had depended on this trade left the heath and moved into other areas of work. By the mid-twentieth century British heathland was effectively 'out of production'.

Grazing

Foremost among the many uses of heathland was its value as rough grazing for livestock. Throughout history heaths had been used as areas of essential additional pasture for a range of beasts. Most heaths were regularly grazed, and could be littered with animals at certain times, especially in spring, when animals such as oxen and draught horses would be turned onto the heaths after their ploughing duties were fulfilled. Spring and summer grazing was the norm, with the animals brought under cover during the more inclement months. However, on some British heaths livestock was left out virtually all year, especially if the owner had no access to better pasture. The constant pressure of grazing kept trees and shrubs at bay, maintaining the heathland environment, and yet it was important to ensure that areas were not overgrazed; grazing was therefore subject to highly regulated rights of access which had developed as a result of usage, ownership and custom (see p.44). Such rights still survive today on some heaths – most notably in the New Forest – and in certain places are being revived as a way of maintaining the heathland landscape.

Tending to livestock on the heath was a time-consuming task, for although the animals could be left to their own devices most of the time, the generally unfenced nature of many heaths meant that some beasts were liable to stray. Pounds – in which escapees were placed to be claimed later by their owners – were therefore a regular feature on the edges of many heaths. Furthermore, the lawless nature of some of the more remote heaths meant that sheep-stealing and cattle-rustling were always a concern, and that a wary eye was called for on the part of the shepherd or cowherd. Many heaths were also visited periodically by herds or flocks being driven on their way to and from market, a process which involved the movement of thousands of animals on a regular basis.

PREVIOUS PAGE **Making a living on the heath was a constant struggle. Heathland folk led a tough life, as depicted in this 1881 painting *Women on the Moor* by Danish artist H.N. Hansen.**

Until the Second World War livestock were a common sight on most heaths. Those left out all year round would seek shelter in winter in the denser gorse brakes, and, generally speaking, there was adequate forage and grazing to keep them going till spring. This photograph was taken on a Dorset heath *c.*1899.

When on a heath most animals will graze primarily on grasses, at least when given a choice. Purple moor-grass, for example, can make up to 75 per cent of the diet of a horse or cow, at least when available in adequate quantities. Sheep will eat sheep's fescue, not surprisingly, which although not as nutritious as purple moor-grass certainly comes into its own when the latter dies back in early autumn. Other plants are also attractive to grazing animals. Both cattle and horses, for example, will eat ling, but tend to shun cross-leaved heath or bell heather. Horses and goats relish young gorse, and goats are also enthusiastic about tackling tree saplings such as birch, oak and Scots pine. In order to generate new grass growth, some heaths were traditionally burned annually, although this practice was not as general as previously believed. Indeed, there are historical documents detailing the punishments meted out to those who did fire the heaths irresponsibly. However, it was certainly regular practice on the heaths and commons of south-west Pembrokeshire, for example, where as recently as sixty years ago young boys would be given boxes of matches by their fathers in the autumn, usually soon after the first frost, and told to go and 'burn the common'!

Heather and turf-cutting

Not surprisingly, heather looms large on the list of heathland products. It was used for a range of functions, of which the most fundamental was as fuel. Heather was traditionally cut in turves, a turf being a square of the matted root material and ground matter, measuring roughly nine inches by eighteen (22.5 x 45 cm) and usually still carrying its embedded heather plants. The depth to which it was cut varied, but could never be less than three-quarters of an inch (2.5 cm) – otherwise it fell apart – and was usually considerably thicker than this. The common right of turbary (see p.44) was a valued commodity, and in areas like the New Forest most holdings held such rights, usually specified to a precise number of turves to be taken in any one year. Forest authority Colin Tubbs suggests that individual annual rights in the Forest averaged about 4,000 turves and gives a figure of 1,500 qualifying dwellings in 1858, for example; this adds up to a potentially vast total of six million turves cut every year.

Turf was usually cut in summer, using a special spade. This was a carefully monitored process; the 1810 Court Books of Moreton Manor in Dorset refer to fines being levied if quotas were exceeded or inappropriate equipment used: 'such furze and turf shall not be cut with any instrument other than a spade or hook under the penalty of ten shillings each offender'. Areas of young heather growth in damp or wet areas were most favoured, as they were both easier to cut and held their shape better than older growth. Turves were cut in strict rotation, to ensure that grazing value was not lost and that regeneration was adequate to ensure a continued supply. A chequerboard pattern was

The cutting of turves was an integral part of heathland life, and in areas such as the New Forest it was a highly regulated affair. This drawing was made by Arts and Crafts artist and Forest resident Heywood Sumner, and appeared in his *The Book of Gorley* (Chiswick Press, 1910).

followed, with two turves left for every one removed, and after seven years or so the heather would usually have grown back enough to be cut again. Once cut, the turves were stacked on end and left to dry on site, and were then removed and stored in a dry, well-ventilated space until use. Turves do not give out a high heat, but they burn well and long, and were the standard fuel for many heathland inhabitants until the advent of cheap coal and paraffin. Indeed, such was their efficacy that they were equally popular among town inhabitants: in the eighteenth century the Dorset port of Swanage was one of the main markets for turves from the Purbeck heaths, for example. In most parts of Britain the practice of cutting turves had all but died out by the Second World War, but it persisted until fairly recently in parts of Ireland. Off the Antrim coast on Rathlin Island, for example, the scars on the heath and the characteristic drying cairns can still be seen today.

Turf-cutting was a real heathland industry. Not only were the turves used for domestic fuel, but they provided an energy source to power the other industries that were found on or near the heaths. For example, as early as the seventeenth century turves were used to fire the furnaces of the Earl of Huntingdon's copperas and alum works at Poole in Dorset. In fact, the Earl tried to enclose the heath from which the turves were cut (and over which common rights had been enjoyed) but had his fences torn down by the locals and then lost his case against them in court. Turves were also used for roofing and construction (see p.78) and as a foundation material in road-building. On some of the continental European heaths, especially in Denmark, turves were used in order to absorb the excrement of animals in their byres and were then spread across arable land as a fertiliser, a system known as *plaggen*. It is not certain whether this practice was ever followed in Britain.

Heather was also used as fodder for animals, as low-grade thatch, for stuffing mattresses, in the manufacture of brooms (see below), in honey production (also below) and for the making of ale. As a dyestuff it yielded a yellow-brown dye, which could be rendered more olive in hue if oak galls were added or if it were boiled in an iron pot. Heather also had various medicinal uses, mostly connected to its antiseptic properties. For example, the young shoots were boiled up to produce a decoction used both as a cleanser and to bathe wounds; it is also long known as being of special value to those with kidney and urinary tract problems.

Heather was traditionally managed and controlled by cutting and burning. However, it is also subject to attack by the heather beetle (see p.106), which can cause extensive damage to healthy plants.

Gorse

Gorse or furze was one of the most highly prized products of the heath, of vital importance to those who lived there. Its main value was as fuel and fodder, but it was also called upon to meet a range of other requirements from sweeping chimneys through use as a thatching material to fencing and the making of dye (from the yellow flowers). However, gorse was paramount as a fuel material. Indeed, its collection was included under the common right of estovers (see p.44), and the sight of 'furze-hackers', often children, gathering gorse faggots (or bundles), securing them with lengths of bramble or briar and then carrying them home, was a classic heathland scene until the early years of the twentieth century. The harvesting of gorse was carried out throughout the year, although younger gorse was most abundant in spring and summer; by winter, it was mostly older growth that was cut. Some areas even celebrated the furze harvest; at Meneage in Cornwall, in a tradition known as 'leading furze', a horse would be led through the village with gorse faggots on its back, signalling 'a season of jollity and mirth'.

Gorse was usually cut using a sickle and billhook, both of which were slightly adapted to suit the particular characteristics of the plant, and to guard against the spines hands were often protected by gloves or by rope wound round the fingers. Being a 'faggoter' could be a full-time occupation, for faggots were gathered *en masse* on some heaths and then taken into the towns for sale at market, where demand was great. There were even regulations about the size of a faggot, and rules about where in a town they could be stored, due to the fire risk. There was money to be made too; Chris Howkins estimates that 'a good faggoter could cut and tie a hundred faggots a day and for a long time in recent history that day's work would have earned him 2/6d ... '.

Both the thorny twigs and woody stems of the plant were collected. The former were popular for use in ovens (especially those of bakers), owing to their tendency to burn very quickly and strongly, creating instant heat. The older wood gave out a longer-lasting heat and so was more favoured for the firing of tiles and bricks and for lime kilns. Gorse was also widely used in the home, although the preferred domestic fuel was often the slower-burning turves (see above); gorse was always a useful kindling material, however. Even so, gorse – as with turves – fell out of fashion in the face of the arrival of cheaper, alternative fuel sources some one hundred years ago.

As far as livestock was concerned, gorse was exceedingly valuable as shelter in the winter and also served as a very important fodder for horses and, to a lesser degree, cattle. In order to boost the nutrient content, it was often mixed with other material, such as barley meal. Young gorse (especially of the western species, *Ulex gallii*) was most favoured, having less formidable spines. Older growth was crushed before being fed to the animals, either by beating with stones or mallets, or through processing in a gorse-mill. Gorse-mills date back to at least the mid-eighteenth century, and the earliest were powered by water; they housed a free-standing rack in which a central shaft or cylinder with spikes would be driven, crushing the gorse as it was fed into the machinery and spewing out the chopped pieces below. They were particularly common in Wales, and one of the best examples is now in the Museum of Welsh Life at St Fagan's, Cardiff. In

This 1799 coloured engraving of a furze-cutter offers a rare contemporary insight into what was once a common rural occupation. The gorse sprays can be seen piled up in the background.

In order to render them more palatable to livestock, gorse cuttings were usually crushed in some way to break down the thorns. The large boulder in this picture was used for precisely this purpose on a Pembrokeshire farm until the early years of the twentieth century: smaller stones were used to pulverise the cuttings against the boulder, now abandoned under rubbish in a barn.

later years gorse was processed into fodder by hand-turned portable gorse-mills, and more recently by diesel-powered chaff-cutters – still a regular farmyard activity in Pembrokeshire at least until the Second World War. It could also be mashed using a special implement with a long handle and two metal blades crossing at the end, which was used to pound the gorse in a stone trough or half-tub. In the New Forest this implement was called a 'fuzz-chopper', but doubtless many other regional terms exist.

Gorse-mills such as this example were powered by water. The axle on the right was coupled to the waterwheel and drove the central shaft; this then turned between the fixed spokes, thereby mashing the gorse (which would be fed into the machine by hand).

Demand for gorse was such that what grew naturally on the heath was usually insufficient. Considerable effort therefore went into managing it as a renewable resource, to the extent that it was also actively cultivated as a field crop, a process that spawned many hundreds of 'temporary heaths'. Idle land was often planted up with gorse seed bought from commercial suppliers, and in parts of nineteenth-century Wales, for example, virtually every farm maintained a few areas of cultivated gorse to ensure an adequate supply of winter fodder. 'Gorse sales' were regular events, and high prices were paid for prime cuttings. Today, however, the use of gorse as either a fuel or as animal fodder is defunct, alive only in the memories of the elderly and in abandoned machinery and tools consigned to attics and derelict barns.

Bracken

Nowadays much maligned for its invasive qualities, bracken was historically a much-valued heathland plant and one of considerable economic importance. Its main use was as bedding material, primarily for livestock but also for humans, but it also served as a basic compost, being spread over fields as a surface mulch and then ploughed in, and as a particularly effective fuel, for it burns quickly and fiercely. There is evidence from Farnham in Surrey that during the Middle Ages it was even used to fire tile and brick kilns in the town. It was equally popular as a packing material, to safeguard the transportation of china, glass and other breakables, and its absorbent quality also made it highly suitable for use as a floor-covering, both within the home and outside, in muddy farmyards and on tracks and roadways. Perhaps the most lucrative use of all, however, was the burning of bracken to produce potash, an effective fertiliser and one of the essential ingredients in both glass- and soap-making. Huge bonfires would be made of it, comprising a cartload or more and often taking several days to burn through, revealing the potash below. Bracken was therefore a truly multi-purpose product, and its harvesting, sale and use were once a mainstay of the heathland economy. It should, however, be remembered that bracken was not found on all heaths, being essentially a forest plant, albeit one with great opportunistic tendencies. It is often most abundant on heaths that are directly adjacent to woodland, such as in the New Forest.

Gathering bracken on the heath at Duddleswell in Ashdown Forest, 1897.

Bracken was usually cut green, in summer or in early autumn, when the fronds were at their greatest extent and before they started dying back at the onset of the first serious frosts. It was either cut with a scythe or knife or pulled by hand, and then piled up on carts to be taken away for drying in large stacks known as 'bracken-cocks'. Such was bracken's value that there were restrictions on when and where it could be cut, and bracken stands were often the scene of conflict between contending would-be harvesters: in Surrey, Chris Howkins refers to there being, within living memory, counterclaims and disputes between families over the right to cut a particular area of bracken. However, the commercial cutting and use of bracken has all but died out in the last five decades or so, its removal now more an issue for conservation managers than anyone else, although research continues into its potential use as a viable commercial compost.

Broom and broom-making

Despite being somewhat toxic, broom was valued by heathland dwellers for a range of medicinal uses (it is a powerful diuretic, for example) and also in food and drink; the young buds were pickled and salted and may even have been a sort of delicacy, to be 'used for salads as capers be and eaten with no less delight', if the herbalist Gerard is to be believed. However, the main value of broom was, as the name suggests, in the making of that invaluable domestic item, the broom or besom.

The making of brooms was a common and widespread activity on many heathlands, but especially on those in Surrey, the home of the legendary 'broom-squires'. The romanticism that has grown up around these people – partly because of the fanciful but entertaining novel *The Broom-Squire* (1913), written by S. Baring-Gould and set on the heaths around Hindhead and the Devil's Punch Bowl – belies what was undoubtedly a hard and poverty-stricken existence. Brooms were made from three main heathland products:

heather, broom and birch. Although the popularity of heather and broom declined in favour of brooms made from the more durable birch, the technique of assembly was essentially the same for all three, insofar as the sweeping material or 'spray' was gathered in a bundle or 'head' and bound tightly round a handle or 'tail'. The latter was usually fashioned from birch or hazel poles, although other woods such as ash, sycamore and lime were also used; the head was secured by 'bonds' made either of bramble fronds or of strips of bark or wood, mostly later replaced by wire.

The town of Verwood, close to Bournemouth, was home to various industries associated with the surrounding heathland. These included pottery, for which the town became famous, and broom-making, shown here.

Chris Howkins has highlighted how broom-making was a real cottage industry, with whole families involved in various parts of the process, much of which took place immediately in and around the home. The birch stems were generally cut by the man of the house, usually from coppice stools, in winter and early spring, and were then taken home, where his wife would sort and trim them to the appropriate length (generally reckoned to be three feet [one metre] or so). The stems were then gathered into faggot bundles and stacked, ready for use; when the time came to make the brooms, the bundles were broken open and separated into appropriate widths: a diameter of ten to twelve inches (25–30 cm) was normal. Meanwhile, the children would soak the bonds in water to make them pliable. The making of a broom was very much man's work – hence the term broom-squire – and was carried out over a wooden horse, made specially for the purpose. Howkins describes the technique:

> The horse had a clamp that was opened and closed by a foot lever and into this clamp was held the end of the bond while under it, and across the horse was gathered the spray. The bundle was gathered tightly and then pulled even tighter by rotation against the bond to bind it round. This was done three times for birch but only twice for the smaller ling brooms.

Once enough brooms had been made to merit a sales trip, the husband would set off in a cart or wagon, touring local markets and fairs to sell his wares. Often a range of brooms would be made, of different sizes and materials, for although most brooms were of birch, those made of heather ('ling besoms') remained popular for indoor use, being rather softer. One of the British headquarters for broom-making was the village of Tadley, near Basingstoke in Hampshire. The village once had a London County Council contract to supply 15 gross of brooms at a time, and in its heyday was estimated to produce some 100,000 brooms per annum. British broom-squires (also known as 'broom-dashers') continued to be active until well into the twentieth century; the last surviving one in Hindhead was George Mayes, who lived there until 1939, making brooms, grazing his cattle and selling their milk. He was a true heathland man, only leaving the area once, for a two-week spell in hospital. Exponents of his craft continue to produce brooms today, although as a properly commercial enterprise broom-making of this type is dead, a victim of cheaper, mass-produced alternatives.

Minerals and mining

Heathland soils may be poor in terms of nutrients, but they can render surprising riches to those prepared to dig. 'Common in the soil' – the right to take sand, gravel and minerals – was well established on many heaths, and there are repeated references throughout history to heathland dwellers making their living, at least in part, from this activity. Thomas Hardy's Egdon Heath wizard Conjuror Trendle was 'a dealer in furze, turf, "sharp sand", and other local products', for example. Sand and gravel were usually sold to the building trade for use in construction.

Clay was also a valued commodity, and as 'marl' was often dug for use as a soil improver (see p.47). There were other uses, too. On the Pembrokeshire heaths, for example, clay was mixed with culm (coal dust or slack) and a little water to make a form of domestic fuel. Horses – led round in a circular motion – were used to thoroughly mix the ingredients in a shallow pit and the black, putty-like mixture that resulted was shaped into so-called culm balls, usually by women, and used in grates rather like coal. It gave out a smouldering, low heat and if allowed to die out could be reignited the next morning. Many poorer families relied heavily on it. On Trefeiddan Moor near St David's clay-pits some three metres (up to ten feet) deep have been found, evidence of the former importance of this activity locally. It appears to have died out in the mid-twentieth century.

In Cornwall, the heaths are still the location of a lucrative china clay industry, which came to the rescue of many tin-mining families when that industry collapsed in the face of cheaper imports during the late nineteenth century. 'Tinning' was once a common occupation on Cornish heaths, with the first tinners working out in the open, using a pick and shovel to extract tin from the exposed seams in the granite that littered the landscape. By the mid-fifteenth century it was necessary to dig the first underground shafts in order to continue reaching the tin, and mines were subsequently established in many parts of Cornwall, often in very remote areas. Copper was also mined on the heaths, and by the time of the Industrial Revolution the industry was booming. For the next century or so it was a mainstay of the Cornish economy, with some 50,000 men working in the local mines. Disaster struck, however, from the 1870s onwards, with an increasing flood of cheaper tin and copper from British colonies overseas. Many Cornish mines closed down and thousands were thrown out of work. The industry never really recovered, and the last tin mine (South Crofty) closed in 1998, although at the time of writing (November 2002) it is being pumped out with a view to production resuming during 2003. Meanwhile, the heathland landscapes of north Cornwall and West Penwith are still dotted with the characteristic ruins of the engine-houses and their chimneys, and with the austere stone cottages of the tinners and their families.

Old clay-pits on Dowrog Common in Pembrokeshire. The pits are now a valuable wildlife habitat.

The Cornish china clay industry was established in the eighteenth century following the discovery in 1746 of china clay or kaolin by chemist William Cookworthy on Tregonning Hill. The local china clay is of particularly fine quality and was initially used in the manufacture of porcelain, although subsequently paper overtook ceramics as the prime user of the clay. The industry prospered during the nineteenth century, absorbing many of the workers who were forced out of their jobs down the tin mines, and today there remain over twenty live clay-pits in the county, with others in Devon. Although not all the pits are located on heathland, the impact of clay-mining on the environment is dramatic wherever it may be: the huge, pale mounds of spoil towering above green, water-filled craters at their feet can hardly be overlooked. Yet whilst not all may agree with Daphne du Maurier's description of 'the strange, almost fantastic beauty of the [clay-mining] landscape', given time, the scars do heal. Old pits can be rich in wildlife, and Tregonning Hill, where the heathland has steadily reclaimed the land from which it was stripped, is now an SSSI. Meanwhile, at Caerloggas Downs, an unsightly spoil-heap has been restored to heathland, one of several such projects taking place in the county under the Tomorrow's Heathland Heritage initiative (see p.133).

Many heaths are rich in industrial archaeology. Hartland Moor in Dorset still bears the traces of the railway track on which small steam engines pulled trucks of clay from the pits on the heath.

Clay-mining was also a major activity on the heaths of the Isle of Purbeck in Dorset. The high-quality ball clay was carted across the heathland to wharves on the River Frome and thence taken by barge to porcelain works further afield, which most notably included the Wedgwood factory in Staffordshire. The open-cast pits can still be seen, as can the derelict narrow-gauge railways that replaced the miners' carts and once criss-crossed Hartland Moor, for example. Mining continues in the area today, but the clay now leaves by diesel truck.

On the heaths of Surrey the mining was for ironstone, which was the mainstay of the former Wealden iron industry. The pits from which the ironstone was extracted can still be seen on Thursley Common, where an iron-making works was established in 1610. The industry is long dead, but the legacy of pits and hammer ponds, many of which are now filled with water and plantlife, form an important habitat for invertebrates such as dragonflies.

One of the most interesting heathland mining enterprises took place on the heaths of the Anglian Brecks, which for much of the last two hundred years were renowned for their flint-knapping. 'Knapping' is the term for working a piece of flint into a particular shape or use, and has been known from the Breckland heaths for at least 4,000 years. Initially the flints were fashioned to make axe-heads and knife-blades, but it was for use in flintlock guns that flint came into increasing demand from the seventeenth century onwards. The centre of the trade was the small town of Brandon in Suffolk, and most of the flint was mined on an area of neighbouring heathland known as Lingheath. Flint-mining was an arduous and potentially dangerous occupation, with the miners working singly in the pits, excavating large slabs of flint which were then carted across the heath into the town, where they would be 'quartered' into smaller fragments and finally knapped into the small flints required by the arms industry.

By 1800 Brandon was the sole supplier of flints to the Master General of the Board of Ordnance and thirteen years later there were some 160 knappers working in the town, producing more than one million gunflints per month. Flint was also used in construction, to create decorative chequerboard patterns within stonework tracery (known as flushwork), and was an important component in road-building, where it was valued as a strong base material. Indeed, many local people would supplement their income by 'stone-picking', i.e. by gathering the lumps of flint that littered the surface of the heaths and selling them by the bucket to road-menders. The advent of modern firearms set the Brandon knapping industry into decline, although a stay of execution was provided by the continued demand for gunflints from overseas, right up until the 1950s. Today, a handful of knappers continue to work flint, supplying firearm enthusiasts and decorative pieces for new buildings and restoration projects.

Beekeeping

For those forced to scratch a living from the heath, beekeeping could provide a valuable source of income. This was particularly the case in the New Forest and on the Dorset heaths, where many heathland households kept hives. The main nectar-producing plants on lowland heath are heather and blackberry; gorse and broom are far less attractive to

'Pony' Ashley, the last of the Lingheath flint-miners. This photograph dates from 1931, and Mr Ashley continued to work until the age of 80, a few years later. One local resident recalled seeing him 'walking home ... with a sack on his back and white from top to toe with chalk'.

bees as a source of nectar. Interestingly, in times of nectar famine, bracken – despite its lack of flowers – will also be visited by bees, attracted by the extra-floral nectaries on its leaf stalks. Meanwhile, honey derived from heather is particularly flavoursome and so was traditionally highly prized, commanding a good price in the marketplace. It has a slightly bitter taste and a rich reddish-brown colour. However, the honey produced from *Calluna* is thixotropic or jellified; this makes it difficult to extract, as it needs to be pressed out of the frames. Honey made from *Erica* is more liquid, and a blend of the two was generally considered ideal.

Until the eighteenth century honey was used as an everyday sweetener in Britain. Sugar was imported and therefore expensive (more than twenty times the price of honey at some periods) and honey was an essential commodity in every housewife's larder. It was a particularly vital ingredient in the fermentation of mead and other alcoholic beverages such as cyser (made from apples) and morat (from mulberries). Honey was also used medicinally, for example to treat wounds such as cuts and burns (as it still is today in parts of Africa and the Middle East). Nor was honey the only contribution made by beekeepers to the heathland economy: beeswax was also notable, and in fact a more lucrative product than honey until the Reformation, when the demand from chandlers for beeswax to make candles for the clerical establishment decreased dramatically following the banning of altar candles. Wax was also used in cosmetic creams, for the making of seals on official documents and as a major component in floor and furniture polishes.

In certain parts of Europe bee skeps were traditionally decorated. On the extensive heaths near Lüneburg in Germany, for example, they were often given human faces. In a rather sinister touch, the bees enter and exit via the mouth and eyes.

The main season for heathland honey production was traditionally August and September (the peak flowering season for heather) and the honey would be sold at harvest festivals and autumn fairs. Until the mid-nineteenth century bees were usually kept in conical wicker baskets known as skeps, which were plastered or 'cloomed' with a mixture of lime, mud and cow dung to help exclude light, draughts and rain. As many as 40,000 bees could be accommodated in a single skep, to which they gained access through a small hole cut in the side. Skeps were sometimes decorated, and particularly fine examples are known from the heathlands of Germany, especially in Lower Saxony.

Fruits, berries and flowers

The gathering of wild produce was an essential part of heathland life. During spring, women and children would sell posies of wild flowers by the roadside, but the real period of activity was in late summer and autumn, when the collection of blackberries, bilberries and sloes was an important occupation. With the coming of the railways in the nineteenth century it became easier to send such produce to town markets. Certainly bilberries – also known as whortleberries or hurts – were sent to London from the Surrey heaths for many years, initially destined for the dyeing industry and then latterly as a fruit for use in jams, pies and puddings. They were sometimes packed in punnets made from woven strips of thin birch bark.

However, the majority of the fruits and berries gathered on the heaths were used by the heathland folk themselves, either eaten fresh, or made into jams and preserves, which provided a welcome treat during the long winter months. Home-brewing was also popular, with beers, wines and other concoctions made from a vast array of ingredients, more typically heather and bog myrtle, but also including bog bean (also known as the 'bog hop', and which was still being gathered by gypsies on Hartlebury Common in Worcestershire as late as 1940) and sundew, the bitter taste of which was valued in liqueur-making.

A characteristic heathland species, the linnet was a popular cage-bird during Victorian times. Along with other finch species, as well as larks and buntings, thousands of birds were trapped and offered for sale. For heath-dwellers this could be a valuable source of extra income.

Rabbits

Light, sandy soils are certainly good for digging, and so it is no surprise that rabbits – known as conies until the eighteenth century – have been closely associated with the heathland landscape ever since their introduction to Britain by the Normans (see p.98). Initially they were kept solely for food and usually 'managed' in open, unenclosed areas of heath. However, as the commercial worth of their meat and fur intensified, they were increasingly placed under the careful surveillance of a warrener and his assistants.

Warreners were one of the highest-paid manorial officials and they were often accommodated in purpose-built 'warren lodges', constructed on the highest part of the warren over which they held responsibility. By the fourteenth century rabbit-farming was a significant contributor to the economy of many heaths, and particularly so in the Brecks, where many extensive warrens were established. As the number of rabbits expanded and their value grew, so it became necessary to enclose and protect them more effectively. This was usually done by means of ditches or banks, the latter often built of grass sods topped by gorse faggots. Regular maintenance was required to prevent escapes and stop poaching.

The main period for 'harvesting' the rabbits was from September until March, when numbers were at their highest following a spring and summer breeding season. The animals were hunted down using ferrets, nets and lurcher dogs, a cross between a sheepdog and greyhound. The ferrets were released into the burrows to force the rabbits to the surface, where the burrow exits were netted. The lurchers would chase down any escapees. Rabbits were also caught in traps called 'tip-traps' or 'tipes', pits which were covered by a swivelling iron cover hidden by hay, on which the rabbits would be encouraged to feed. The cover was set to collapse under the weight of the rabbits, which would then fall into the pit below. As many as 2,000 animals could be caught in a single night in this way.

Thetford Warren Lodge was built in the fourteenth century to protect the Prior of Thetford's rabbits. All that remains today is the central tower, the rest of the building having been destroyed by fire in 1935.

Warreners at North Farm, Barnham, Suffolk, photographed on 31 March 1921. The smocks – usually a tawny colour – are characteristic, and the picture is interesting for displaying so well the tools of the warrener's trade: dogs, ferrets (in the boxes) and long-handled staves. The lengths of rope draped over some of the ferret boxes are probably the lines that were fastened to collars around the ferrets' necks when they went underground. In addition, the warreners would have carried nets. The horse and cart used to transport the dead rabbits off the heath can be seen behind the men.

Penalities for poaching rabbits were often severe. In 1805 a man was sentenced to six months' imprisonment and a public whipping at Brandon in Suffolk for stealing two rabbits from nearby Wangford Warren. Transportation and hard labour were also common punishments. Despite this, poaching was widespread and warreners were obliged to set elaborate defences to protect their charges. These ranged from spears and stakes, linked with barbed wire and set a foot or so above the ground to lacerate the legs of the poachers' dogs, to spring-guns and even man-traps. Rabbits were big business, and extensive resources were devoted to their care and protection. In the 1920s, for example, half of the Elveden Estate in Suffolk was given over to rabbits, with thirty warreners killing over 120,000 rabbits per annum.

The effect of all these rabbits on the heathland landscape was very significant indeed. Where numbers were excessive, they could strip the soil of almost all its vegetation, and in extreme cases this overgrazing could lead to a local form of virtual 'desertification' and distinct unpopularity in some circles. For example, in 1549 all the rabbits on a warren at Freckenham in Suffolk were destroyed by angry local villagers following a sandstorm, and a little over a century later the rabbit-induced shifting sands of nearby Lakenheath Warren engulfed the village of Santon Downham and partly blocked the Little Ouse river.

Rabbit farming continued on the heaths around Brandon, for example, until the 1950s. For some farmers rabbits still constituted their main source of income then, being sold either for table or for their fur. The latter was often sold for use in felt hats, the felt being made by the sticking together of the loose rabbit hair with shellac. Although the demand for both rabbit meat and rabbit fur declined steadily during the twentieth century, the *coup de grâce* was the collapse in the rabbit population following the advent of myxomatosis in 1953 (see p.99). In a few months, as the disease swept Britain, one of the most important heathland resources was all but wiped out. The population has since largely recovered, but the rabbit has never regained the commercial value it once held. However, remnants still survive of the days when the rabbit was truly king. Thetford Warren in Norfolk, once one of the most productive heathland rabbit warrens in England, may have been mostly lost beneath forestry but it still boasts the remnants of its once grand lodge, now looked after by English Heritage.

Earning a living on the heath certainly required making the most of what was available, so heathcroppers lived by a combination of most, if not all, of the activities and products detailed above. Diversification and flexibility were key; a few eggs for sale here, some milk or cream there. Meanwhile some heathland residents became adept at operating in surprisingly obscure niches. One such character was 'Brusher' Mills (1840–1905), who used to catch adders on the heaths of the New Forest and sell them to London Zoo for a shilling (five new pence) apiece. Those surplus to requirements were boiled up to make adder fat, used as a traditional medicine. This type of ingenuity was typical of heathland folk, and was required in no less measure when it came to the business of building a house.

A Home on the Heath

Relatively few people made their home on the heath from choice. Water was often scarce, and the poor soils made cultivation difficult. Heathland dwellers were therefore forced to make do with what the heath provided. They were invariably poor, their dwellings modest. Most lived in basic cottages, which in many cases would have been little more than single-storey, one-roomed affairs. The walls were often constructed from stacked heather turves, and the roof either covered with turves (laid flat, abutting each other and sometimes placed over a bed of bracken) or thatched with heather, gorse or bracken stems. Timber was not always available or affordable, in which case the rafters would be made from gorse trunks and branches. This made for narrow houses, as their width was governed by the height of the trunks (rarely more than four metres, or thirteen feet). Salvaged or poor-grade timber was sometimes used to create a ridge and gable ends. Grander houses – although the term is strictly relative in the heathland context – might have a proper timber structure, allowing a two-storey edifice, and walls made of wattle-and-daub panels.

Even in areas where clay was readily obtainable, bricks were rare, being time-consuming to make, and therefore expensive. The clay would be dug from pits and then left in piles over the winter to temper. Brick-making would begin in the spring, but only if the

The National Trust's Oakhurst Cottage, in Hambledon near Godalming in Surrey, in a 1991 painting by Richard Sorrell. The building is typical of the grander type of heathland home, with a timber frame, wattle-and-daub walls and a pantiled roof.

weather was dry; wet conditions meant the bricks would not dry out and so could not be fired. If a householder could only afford a limited number of bricks in an otherwise timber/wattle-and-daub construction, then these were usually reserved for use in the hearth and chimney, or for the gables. Stone was generally unusual, being uncharacteristic of most lowland heath landscapes; it was, however, popular for use in making a hearth and threshold or doorstep. More common as a heathland construction material was marl, a type of clay used in the construction of so-called 'cob houses'. The lumps of marl were heavily moistened, mixed with straw, heather or chopped gorse (to help bind the material), and then beaten or trampled until they could be shaped readily into small building blocks the size of a double fist. The walls were then raised in courses, each course being left to dry out and harden naturally before the next was applied. When complete, the walls were given a lime-wash coat, to prevent water ingress and without which the clay begins to collapse, leading to serious structural problems! With regular limewash protection and maintenance cob buildings can last many years, and thousands can still be seen in the south-western counties of England, especially Devon.

The poorest heathland inhabitants were unable to afford even the most modest permanent home. Instead they often lived in 'benders', a form of tent made from hazel or birch poles covered in sacking. This photograph of an elderly gypsy couple in Ashdown Forest was taken in 1895. The man is weaving a straw bee skep.

Inside, most heathland houses were exceedingly modest. Chris Howkins cites a house containing just one room of eight feet by twelve, in which the occupants presumably ate, slept and carried out all their other domestic activities. Wattle-and-daub walls and screens, or lengths of sacking, were used to sub-divide larger spaces and cover doorways and windows. Floors were of bare earth or sand. Furniture was very basic and limited; the main item in most homes was the bed, described by Howkins as typically 'not a carpenter's masterpiece so much as a heap' with 'a bottom layer of coarse twigs or furze to allow air to circulate and urine to drain through'. A mattress would be made of heather, bundled and stacked, with a softer layer on top of bracken, hay, grass or leaves stuffed into a large sackcloth slip. Conditions were generally less than sanitary, although the situation was often improved through the use of aromatic plant species such as bog myrtle, also useful as an insect repellent, and lady's bedstraw, so-called because of the tradition of using it to stuff the mattresses of ladies (even those of high rank) and especially valued for its astringent, vaguely antiseptic qualities.

Fire was an ever-present danger. Houses were often tinder dry, and one spark from the fire would be enough to engulf a dwelling made of little more than desiccated heather turves, gorse and low-grade timber. Originally, the hearth – the focal point of everyday life – would have been in the centre of the house, the smoke escaping via the doors, windows and thatch, but more latterly most householders were able to afford a stone or brick hearth and chimney at one end of their property. Overall, however, heathland homes were hardly salubrious. They reflected both the unforgiving nature of their landscape and the harsh lives of their inhabitants. As the twentieth century progressed, many were forsaken in favour of better accommodation away from the heath, eventually being condemned as being insanitary. Demolition often followed.

Heath bedstraw (in the foreground) and lady's bedstraw (behind) were both used to stuff mattresses and cushions. Their aroma helped counter what could otherwise be rather overpowering domestic odours.

Chapter Four

Plants and wildlife

PLANTS AND WILDLIFE

> To have always about me that wilderness which I best loved – the rude incult heath, the beautiful desolation; to have harsh furze and ling and bramble and bracken to grow on me, and only wild creatures for visitors and company.
>
> W.H. Hudson, *Hampshire Days* (1903)

A heath in high summer is a magnificent sight. Positively burgeoning with life, the apparent uniformity of tracts of heather, grass and gorse belies a complex series of smaller habitats, each providing a home to a range of other plants and a wealth of invertebrates, reptiles, birds and mammals. Lowland heath supports more priority Biodiversity Action Plan (BAP) species of flora and fauna than any other habitat, with 57 species primarily associated with it and a further 79 recorded as using it. For these reasons it has been given priority habitat status in both the United Kingdom BAP and European Union Habitats Directive.

However, relatively few species are specifically dependent on lowland heath. All the types of wildlife found on heaths are there because the heathland habitat offers the appropriate physical conditions for feeding, breeding and achievement of their life cycle, but in the vast majority of cases these requirements can be met equally successfully – and sometimes even more so – in other types of habitat. Certainly the greatest diversity of plants and wildlife on heathland is found on those sites containing the highest number of smaller habitats in the heathland mosaic, i.e. from bare ground through short turf to longer grass, heather cover, gorse scrub and light woodland.

However, the key point is that heathlands do offer a number of rare and range-restricted species the closest British option to their ideal conditions, and are thereby their best – or in some cases, their only – opportunity to exist in this country. Heathlands are therefore an important contributor to national biodiversity. Equally, there is no doubt that some of the species they support are among the most interesting, beautiful and, sadly, endangered of all British wildlife. They are also among the most famous, in the sense that recent campaigns to save heathland have focused primarily on flagship species such as the Dartford warbler and the sand lizard. These attractive creatures understandably generate effective publicity and engage the public's attention, and the decline in their numbers and distribution through much of the twentieth century mirrored the demise of heathland generally. Yet heathland habitats are also of immense value to a range of smaller and even rarer species, particularly invertebrates.

PREVIOUS PAGE **Thursley Common in Surrey is one of the richest heathland sites in Britain in terms of wildlife. The wide range of micro-habitats, ranging from open areas of sand to wet bog, ensures great diversity.**

Plants

Heaths are not the place to expect a vast number of different plant types. The plant community of most heaths is usually dominated by heathers (see pp.16–20), and is rather species-poor. This is due to the generally impoverished condition of heathland soils and is particularly so in the case of dry heath. However, this lack of variety is more

Dorset's Arne reserve, owned by the Royal Society for the Protection of Birds, is noted in particular for its populations of Dartford warbler and nightjar.

Gorse is a classic heathland species, and essential habitat for the Dartford warbler. It has a range of uses and was traditionally exploited to the full by heath-dwellers.

than made up for by the presence of several interesting and rare species, some of which are restricted to the heathland environment. One other notable point about heathland flora is the general absence of trees; the reasons for this are explained in detail elsewhere, but trees do of course occur on heathland, depending on local conditions and land use. The number of species concerned is relatively small: Scots pine and Corsican pine are more typical of drier heaths, with stands of sallow and alder occurring on wetter heaths. Birch is common throughout, and potentially invasive, as are blackthorn, hawthorn and that notorious introduction, *Rhododendron ponticum*. Oak also occurs regularly on lowland heaths throughout Britain, but none of these species can be argued to be associated with heathland particularly. Rather, they are remnants – or the vanguard – of different types of landscape in the chain of succession.

One of the most familiar and ubiquitous heathland plants is gorse, traditionally known as furze or whin. The term 'furze' (or 'fuzz') comes from old country words for any plant with spines or needles – hence 'firs' for conifers – and is most common in southern England. Gorse was also commonly known as whin, a name carried here from Scandinavia and which still persists in northern England, Scotland and Ireland. Where present in large quantities gorse makes for a dramatic heathland landscape, its rich coconut scent attracting a host of buzzing insects and its vibrant yellow flowers a perfect foil to the mellow purple of the heather.

There are three species of gorse in Britain: common gorse (*Ulex europaeus*), the most widespread and abundant of the group, and two more locally distributed species, *Ulex gallii* or western gorse, and *Ulex minor* or dwarf gorse. Common gorse is found throughout Britain and Ireland, whereas western gorse shows a decidedly maritime bias in its range, being found across south-west England, Wales, northern England and Northern Ireland, with an outlying pocket in coastal East Anglia. Dwarf gorse is restricted to southern and eastern England, where it largely replaces western gorse. All three species are fast-growing and occur readily on heathlands; they are, indeed, closely associated with the habitat. Yet they have slightly different preferences as to precisely where in the heathland envelope they prefer to be. Both western and dwarf gorse prosper on the open heath, out amongst heather and often in exposed locations (such as clifftops, in the case of western gorse), whereas common gorse prefers heathland edges. The latter grows taller than its relatives – sometimes up to three metres (ten feet) or so – and is usually more woody, often assuming the form of a small tree. Both the other species are generally less robust and more compact, especially in the case of dwarf gorse, which is usually almost prostrate and keeps well down among the heather. However, there is a further, more telling, difference between common gorse and the other two species: flowering time. Common gorse blooms mainly in spring, from February to June, whereas the western and dwarf types both flower in late summer and autumn, between July and November. On those heaths where common gorse is present with one or more of the other species – a common enough scenario – gorse is therefore effectively in flower for much of the year, hence the origin and value of sayings such as 'When gorse is in blossom, kissing's in season' and 'I'll pay my bills when the gorse is out of flower'.

For such a tough-looking plant, gorse is surprisingly vulnerable to cold weather. Severe winters can have a serious impact on gorse stands, and high winds can be equally damaging, especially on exposed maritime heaths, where the plants can be 'burned off' by gales. Growth can be equally inhibited by the effects of salt spray. The species soon recovers, however, and like other members of the pea family thrives in poor soils and on disturbed ground. Livestock and, to a lesser extent, deer will graze the young growth before the spines become too unpalatable, and rabbits certainly browse extensively on young gorse. However, if left ungrazed or unmanaged gorse will spread rapidly. It offers welcome shelter to a host of insects and birds, particularly in harsh winter conditions, and historically was one of the most significant heathland plants, playing a central role in the heathland economy (see pp.64–7).

Another member of the pea family, broom is similar to gorse in some ways, although it lacks spines and is a rather more open shrub, both in structure and habit, rarely forming the dense clumps typical of gorse. Its golden, vanilla-scented flowers emerge in spring and are highly attractive to insects. Noted for its great powers of germination, it can rapidly colonise burnt areas of ground but is not a great lover of acid soils, and so is found more commonly on those heaths with a calcareous component, such as the Breckland heaths, for example. An essentially southern species, it is prone to frost and

prefers warm and dry conditions, growing readily on the coast where sandy soils prevail; it was once often used to help stabilise dunes and to boost nutrient levels on marginal farmland – like gorse, it has nodules on its roots containing nitrogen, which it releases into the generally poor surrounding soil. Although it can be poisonous, broom was traditionally valued for medicinal purposes and widely used for a range of other everyday functions, most notably – and predictably – broom-making (see pp.68–70).

Bracken is nowadays regarded as the villain of the piece by many countryside managers. Essentially a woodland plant, it is common in many parts of Britain and its ability to resist control measures and then invade and choke out other types of vegetation, especially heather on heathland, has made it both unliked and unvalued. Yet it is not without a beauty of its own, and certainly its elegant fronds can provide both atmosphere and colour at whatever season. They are arguably at their best in autumn, turning a rich russet hue and curling in on themselves at the first frosts. In the past bracken was greatly valued by country dwellers and harvested like any crop (see pp.67–8) but in recent years this practice has all but died out and so, unchecked, bracken has mounted its offensive across the landscape.

Flowering broom, with a male whinchat. Far more localised in distribution than gorse, broom is vulnerable to cold weather and prefers open, sunny sites. It was probably more widespread in the past, and may be helped by global warming to expand again.

An outstanding survivor, bracken can grow up to three metres (ten feet) high in one season and has a highly efficient system of underground rhizomes which store food and allow it to outcompete other flora. It is flexible in where it will grow, thriving equally in open forests, across rough upland pasture and on heathland, where it can have a serious impact on less aggressive plant species. One of the main problems is the resistance of dead bracken to decomposition. The water-resistant litter or straw which accumulates each autumn when the plants die back can pile up to a depth of 40 or 50 centimetres (16–20 inches), preventing the growth next spring of anything other than young bracken, for which it acts as an excellent nursery! However, bracken (and its straw) is not entirely without wildlife value; it can offer valuable shelter to reptiles, deer and some ground-nesting birds, and when not entirely dominant, can provide useful dappled shade to certain plants such as violets, although these will eventually be suffocated out if the bracken remains unmanaged.

Bracken is highly invasive and if left uncontrolled can choke out other vegetation, including heather.

Heath spotted orchids, with tormentil. Wet heaths often support large colonies of this orchid, which is surprisingly variable in terms of petal colour and leaf pattern.

Most other heathland plants are decidedly smaller and less epic in terms of impact. On dry heaths, the flora is often dominated by species of bedstraw and hawkweed, sheep's sorrel, tormentil, slender St John's wort, milkwort and lousewort. Certain species of orchid can also be present. Other plants include mossy stonecrop, which favours patches of bare ground, viper's bugloss, with electric blue flowers that become almost luminescent at dusk, and dodder, a parasitic species whose pink stems can be found entangled around and over its two main heathland hosts, gorse and ling. Common grasses on dry heath include sheep's fescue, bristle bent (mainly in south-west England) and wavy hair-grass, so called because of its hair-like flower plumes.

Where conditions are wetter, a greater number of plant species occurs. Notable are the carnivorous, ground-hugging sundews, their distinctive red-fringed leaves awaiting visits from unsuspecting insects, and two of wet heath's particular jewels, bog asphodel and marsh gentian. The yellow spikes of bog asphodel are a dramatic sight when flowering in late summer, but the plant is equally spectacular out of season, when its stems take on a tawny-orange hue. It is often found in large colonies and is widely distributed across much of Britain, although rarer in the south and east. The marsh gentian favours equally damp areas and also flowers in August to September, but is rather more erratic in appearance, often vanishing for a few years before suddenly reappearing in good numbers. The ability to survive as a dormant bulb underground has undoubtedly helped it on sites where conditions became unsuitable but which have now been improved through management. It is locally common on southern English heaths. Much rarer is the bog orchid, greatly declined because of drainage and now restricted to relatively few sites, notably in Dorset and Hampshire. It is usually found growing among the bog mosses (*Sphagnum*) in valley mires and is often overlooked because of its small size and pale flowers.

Wet heaths host a particularly rich variety of grasses and sedges. The dominant grass is often purple moor-grass, which can develop dense tussocks and, if ungrazed, can become dominant at the expense of less robust species. Cotton grass can also be common, its white tufted plumes a distinctive feature on marshy ground, and other typical species include black bog-rush, pill sedge and green-ribbed sedge. Some of these are also characteristic of upland areas, and their presence on southern heaths is a reminder of the close interface between the soil conditions of lowland mires and upland bogs. Bog myrtle and cranberry are two further species which, although more readily associated with the upland peat of northern Britain and Ireland, are locally common on wet heaths in the southern lowlands.

Certain heaths are outstanding for the rare plant species they support. Two areas are noteworthy in this respect: the Lizard peninsula in Cornwall, much of which is in the care of the National Trust, and the Brecks of Norfolk and Suffolk. Not only is the Lizard celebrated for the dominance in its heather community of Cornish heath (see p.19), but it is also home to over ten per cent of the 200-plus species listed in the British Plant Red Data Book. These specialities include several rare species of clover, hairy greenweed, dwarf rush, fringed rupturewort and land quillwort, the last two being found nowhere

Bog asphodel, with cotton grass behind and sundew in the foreground. The bog asphodel's Latin name *Narthecium ossifragum* – 'bone-breaker' – refers to the traditional belief that, if eaten by livestock, the plant would give the animals brittle bones. Any such effect would, however, be due to the calcium-deficient soil on which the animals were grazing, rather than to any properties within the plant itself.

The lowland heaths of East Anglia in particular are an excellent habitat for a range of mosses and lichens, here seen growing alongside heather.

else in Britain. An impressive supporting cast features most of the species typical of maritime heath, such as sea campion, thrift and spring squill, which in May and June combine to produce a bewildering display of colour and variety. What makes the Lizard so special botanically is the unique geological and climatic conditions that apply there; similarly special – but different – circumstances apply in the Anglian Brecks, where the emphasis is very much on plants requiring hot, dry conditions. These include species more typical of Continental Europe, such as spiked speedwell, Spanish catchfly and field southernwood, as well as plants which, although more at home on the coast, can also thrive on the inland Breckland dunes: sand sedge and sand cat's tail, for example.

The Brecks are also home to some of Britain's most important areas of lichen heath. Lichens occur on heathland wherever they are given the opportunity to do so, i.e. when ground flora is so sparse as to give lichens the space to prosper. The typical species are members of the *Cladonia* family, the so-called 'reindeer' lichens which send up stalks from a dense platform on the ground and bear fruit. Mosses are also a feature of both dry and wet heaths, often carpeting the space between heather and other shrubs, and in the case of the *Sphagnum* bog mosses on wet heath, constituting a rich habitat for other plant species and many invertebrates.

Birds

With heathland birds, the emphasis is very much on quality rather than quantity. At certain times of year – particularly during autumn and winter – most heaths can appear virtually devoid of birdlife, with just the occasional passing crow to relieve the barren landscape. Even during spring and summer the overall total of bird species found on heathland is surprisingly small, both their variety and number constrained by the rigours of the landscape and by associated limitations on food supply and nesting sites. A total of 35–40 species is a healthy one for most heaths. Yet among the birds that do make their home here are some real treasures, birds for which heathland represents both their primary British habitat and, in some cases, their last refuge. Several are BAP priority species and the subject of careful monitoring and specific habitat management regimes.

Although many of the bird species found on lowland heath will also favour other habitats, such as upland moors or rough grazing, there is one notable exception: the Dartford warbler. This diminutive bird has become an icon for Britain's heathland, its fate inextricably linked with lowland heaths and particularly with gorse, on which the British population is overwhelmingly dependent. Indeed, two of the bird's traditional country names – furze wren and the rather more exotic Provence furzeling – allude to this association. Dartford warblers are both restive and furtive in their habits, constantly on

The Dartford warbler is one of the great conservation successes of recent years. Its numbers have expanded considerably, and it is now being reported from localities where it had not been seen since the nineteenth century.

The woodlark is a great opportunist and will quickly take advantage of manmade habitats, such as new forestry plantations. Although at least partly resident in Britain, it shows a tendency to move off heathland in winter, preferring to forage on farmland.

the move within (and between) gorse stands, but almost always skulking and reticent. The best chance of a good view is usually in early spring, when the handsome male bird will sometimes launch his harsh, scraping song from a prominent gorse sprig before returning to the safety of the bush below.

One of only two resident species of warbler in Britain, the Dartford is essentially a Mediterranean bird, common in the sun-baked *maquis* of southern France and Iberia, and really at the limit of its range in Britain. It is no surprise, therefore, that it is mostly restricted to southern England, its headquarters firmly based on the warm, sandy heaths of Dorset, Surrey and the New Forest. However, in the past it was more widely distributed, with records from many English counties, including Kent, where the species was first described – at Dartford – in 1773. However, heathland loss and fragmentation undoubtedly hastened its demise in many areas. Furthermore, the bird's sedentary nature makes it highly vulnerable to adverse winter weather; the British population crashed to a handful of pairs following the severe winter of 1962/3, but the species has recovered strongly and now numbers some 2,000 pairs spread across much of southern England from Cornwall to Norfolk. Although much of this resurgence is the result of a recent run of mild winters, there is clear evidence that increased heathland protection and more informed thinking on how gorse should be managed to benefit Dartford warblers are both playing a significant role. The highest densities of birds occur where the gorse has a compact, dense structure, and whilst the species can tolerate the presence of young tree scrub, when the latter exceeds 40 per cent or so of the ground area then conditions become less favourable.

Another classic heathland species – and one that is also doing well currently – is the woodlark. The fluty notes of this species, often regarded as a superior songster to the better-known skylark, can now be heard in more parts of Britain than has been the case for at least 150 years. Despite its name, this is very much a bird of open country, and yet whilst the population has undoubtedly expanded in recent decades, the species has become increasingly restricted to lowland heaths as its other preferred habitats – semi-natural grassland and downland – have disappeared. Interestingly, woodlarks can do very well within recently planted conifer plantations, many of which have been established on heathland, as in the East Anglian Brecks. Here several hundred pairs of woodlarks may be found, but their continued presence is dependent on careful management. Woodlarks prefer dry habitats with a combination of short and scant turf, together with a scattering of tussocks and small shrubs. Clear-fell areas are ideal, at least for the few years immediately after clearance and replanting, but these soon become unsuitable as the young trees grow up or as the ground vegetation becomes rank.

Two of the most enigmatic and mysterious birds associated with heathland are both essentially nocturnal. The nightjar is perhaps best known for the male's extraordinary song, a monotonous yet evocative drone known as churring and usually delivered from a prominent dead branch or bushtop. Careful analysis of recordings has revealed the nightjar's song to contain up to 1,900 individual notes per minute. Nor is this the only notable feature of the male's territorial behaviour: male birds will patrol their 'patch' in a dancing, jerky flight, flashing their prominent white wing patches (both to ward off other males and to attract potential mates) and often bringing their wings into contact behind their backs to produce a loud 'clap'. Much folklore surrounds this species, not least the contention that the bird drinks goat's milk, hence its traditional country name of goatsucker. As with many such terms, however, the reality is disappointingly prosaic. Nightjars feed primarily on night-flying insects, especially moths, and so are attracted to places where these are plentiful – such as near livestock. Birds will range quite widely on their nightly forays, and have even been known to enter villages and hawk the moths and beetles attracted by street-lights.

One of the latest summer migrants to arrive in Britain, nightjars are seldom on their breeding territories much before mid-May. The nest is a bare scrape on the ground, in which two eggs are laid, and so nightjars do best where there are plentiful open areas; clear-fell is often ideal, and so – as with the woodlark – nightjars have prospered in and around new conifer plantations on heathland soils. Improvements in the quality of existing heathland have also helped, and this habitat now holds the overwhelming percentage of Britain's 4,000 pairs.

The stone curlew is another bird of the night, although it is perhaps more accurately crepuscular, i.e. predominantly active at dawn and dusk, when its eerie, wailing cry is most often heard. Although not restricted to heathland (the classic heathland vegetation of heather and/or gorse does not well suit its requirements), as a ground-nester it traditionally prospered on the grassy heaths and warrens of East Anglia. Agricultural

Stonechats are widely but thinly distributed throughout Britain. However, many suitable areas of habitat remain uncolonised, and overall numbers are showing signs of continued decline.

encroachment, disturbance and loss of habitat to forestry sent numbers into freefall, and it is only in recent years that concerted efforts to secure the future of the species have begun to pay dividends. Some 250 pairs now breed regularly in England, most in specially managed areas of the Brecks (where the vast majority now actually nest on the farmland that has replaced much of the former heath) and on the downs of Wiltshire and Hampshire.

Among the smaller birds to be found on heathland, both the linnet and the yellowhammer are common inhabitants and, although much declined in recent decades, remain widespread. Two further pairs of related species are also worthy of special note: the whinchat and the stonechat, and the tree pipit and meadow pipit. Sadly, the current fortunes of the two chats are giving cause for concern. One hundred years ago the whinchat was a common bird across much of Britain and closely associated with lowland heath, a status indicated by historical names such as gorse chat and furr-chuck. Yet today it is a scarce breeding bird in lowland Britain and continuing to decline. Only in the uplands of northern and western Britain does it appear to be holding its own. Although agricultural improvement and habitat destruction have not helped, its demise in the south and east may be related more to climate change: it prefers moister conditions, such as those found further north and west. Equally, as a migratory species it may have been adversely affected by the increased desertification of the African Sahel. The stonechat, by contrast, is a resident bird, but consequently prone to losses in harsh winters. Once a classic heathland species across southern Britain, it was hard hit by the loss of suitable habitat during the twentieth century and is now much more scarce. It remains, however, a characteristic bird on the Surrey heaths (such as at the National Trust's Witley Common) and in the New Forest, for example, although it is generally slow to recolonise areas from which it has vanished. Stonechats are, however, one of the easiest heathland birds to spot: both male and female birds often perch sentinel-like on the top of gorse bushes or heather clumps, uttering their 'chack chack' alarm call, which closely resembles two stones being clicked together.

Both meadow and tree pipit are fairly common heathland birds, but with slightly different requirements. Although both are ground-nesters, the meadow pipit prefers a more open landscape to that favoured by its close relative, which is able to take advantage of modern forestry techniques by exploiting young plantations. Both species have an attractive 'parachute' display flight, the tree pipit descending to the top of a tree or shrub, and the meadow falling slowly to the ground or to a small bush. The tree pipit can also live successfully in areas of scrub, especially birch, which also provide refuge for other birds more typical of open woodland, such as warblers and tits.

Shrikes are avid predators, feeding on insects as well as small birds and mammals. Food that is excess to immediate requirements is impaled on a thorn bush, to act as a 'larder' for when the need arises. The great grey shrike, depicted here, is now an uncommon winter migrant to British heaths.

Britain's more common birds of prey, such as kestrel and sparrowhawk, will readily hunt over heathland in search of small rodents and birds, but there are also one or two predators more specifically found on lowland heath. Foremost among these is the hobby (see p.136), an elegant falcon which often feeds on the dragonflies attracted to boggy heathland pools. It is one of the fastest birds in the air, capable of snatching a swift or a bat in mid-flight, and readily nests in the isolated stands of pine that are such a classic feature of southern English heathland. The number of hobbies in Britain is increasing rapidly and there are possibly more pairs breeding here now than for several hundred years. Being migratory, they move south in autumn to Africa and are replaced on British heathland by winter predators from the north, such as short-eared owl, merlin and hen harrier; some tracts of heath, such as Roydon Common in Norfolk, have held impressive roosting numbers of the latter in the past. Equally, until the last few decades, heathland was an important habitat for Britain's most regular two species of butcher bird or shrike: red-backed and great grey. Sadly, the former is now extinct as a regular breeding bird here, and the latter – once a frequent winter visitor to many heaths – is now a scarce sight indeed.

Finally, it is important to bear in mind the value to wading birds of lowland heath when it contains areas of wet bog and mire. Curlew, redshank, lapwing and snipe are all attracted to such habitat, both for feeding and nesting, and can do well when conditions are right and disturbance minimal. The New Forest holds significant numbers of all four species on its heathland.

Mammals

Heathland *per se* has limited appeal to most mammals and, with one notable exception, few are found there at any great density. In the case of herbivores this is mainly because there is generally insufficient cover and forage, although deer will use heaths readily when there is adequate scrub on which to browse and in which to rest. Indeed, deer may affect the heathland landscape in the sense that when present in large numbers they can inhibit the encroachment of scrub through browsing pressure. However, no mammals are wholly dependent on heaths and, like deer, most other species will use heathland in conjunction with neighbouring habitats such as woodland and pasture. These include badger, fox, stoat, weasel and small mammals such as voles and shrews, as well as bats. The value of heathland to bats is something that has been increasingly recognised in recent years and it is clear that many bats use heathland habitats as hunting grounds, attracted by the large numbers of flying invertebrates found there, particularly moths. At Budby Heath in Nottinghamshire, for example, five species of bat have been recorded, including the scarce Leisler's. However, bats need holes in which to roost and breed, and the general lack of trees on heathland can be an inhibiting factor. The erection of batboxes can therefore provide valuable and permanent refuges for bats on heaths.

The mammal most closely associated with heaths, however, is the rabbit. Introduced to Britain by the Normans in the twelfth century, rabbits were initially 'farmed' in warrens (see pp.75–8), but soon escaped and established themselves widely. A high reproductive capacity (three to six litters a year, with three to twelve young per litter) helped ensure that they soon reached epidemic proportions in many areas, and their catholic choice of food – ranging from tree bark and buds to germinating cereals and root vegetables – meant that they rapidly became a serious agricultural pest. The difficulty in completely eliminating them, even on a very local basis, as well as the continued demand for their meat and fur, helped ensure that they were allowed to continue as a widespread and common countryside resident.

Rabbits became particularly abundant on grassy heathland, where they had a significant impact on the condition and appearance of the landscape. In many places the soft, sandy soil in which they preferred to burrow became seriously undermined, and there are many instances of it degenerating into areas of virtually bare dune (see p.77). Heavy and constant grazing pressure by the animals also maintained a very close sward, preventing the regeneration of gorse (and other scrub, for that matter), inhibiting the growth and spread of heather and acting as a serious impediment to the survival of some species of specialised heathland flora such as maiden pink and Spanish catchfly. Research suggests that at some sites rabbits were instrumental in turning *Calluna* heath into grass heath dominated by fescue. However, intensive rabbit grazing did create ideal conditions for some bird species, most notably stone curlew, wheatear and, in the Anglian Brecks, ringed plover, all of which prospered on the bare, flinty ground of the rabbit warrens. Equally, there is evidence that intensive grazing by rabbits favours the creation of lichen heath, one of the rarest heathland communities.

By the early 1950s the rabbit population of Britain had reached such a level that farmers, in particular, were clamouring for government action. In 1950 the viral disease myxomatosis had been deliberately introduced from South America to Australia, where the enormous rabbit population had been massively reduced as a result. There is, however, no evidence to suggest that myxomatosis was purposefully brought to Britain, but rather that it arrived accidentally from France. Whatever the origins of its presence here, the impact on the British rabbit population following the arrival of the disease in 1953 was immediate and catastrophic: in most areas numbers were down by over 95 per cent within two or three years, and it is clear that some farmers did deliberately move diseased animals around to try to control rabbits locally.

The decline of the rabbit as a result of myxomatosis was to have a profound impact on the British landscape. The photograph above is of heathland at Murlough in Northern Ireland, and dates from 1955 when the landscape supported a very short grass sward, closely cropped by the large numbers of rabbits. The picture on the left shows the same area in 1994, by now heavily invaded by scrub, the latter unhindered by a greatly reduced rabbit population.

The virtually complete removal of the rabbit population had an extraordinary effect on some areas of heathland. Almost immediately, plants began to appear that had not been seen for decades, and the late 1950s saw the notable and dramatic flowering of some heaths. However, this renaissance was short-lived. Uncontrolled by grazing, more vigorous plant and grass species soon took over, choking out the low-growing specialists and constituting the first stage in the succession to scrub and eventually woodland. Many areas of heath rapidly became rank and overgrown as a direct result of the demise of the local rabbits, and we are still living with the consequences today. For although the British rabbit population has largely recovered from myxomatosis, the relatively short hiatus between their abundance pre-1953 and the subsequent return to something like their previous numbers was a sufficient window of opportunity for advancing scrub encroachment to change totally the character of some heaths.

An excellent example of the impact on heathland of myxomatosis can be seen on National Trust land at Murlough in County Down, Northern Ireland. Designated Ireland's first nature reserve in 1977 and now a Special Area of Conservation, Murlough is an outstanding site for dune heath, of which it holds some twelve per cent of the British total. The invertebrate populations here are particularly notable and the reserve has recorded 21 types of butterfly, including a regular population of marsh fritillary. Dune grassland plants are also significant and feature such notable species as autumn gentian, moonwort and Portland spurge. The dunes were once the site of a vast rabbit warren, and rabbits continued to have a major effect on the landscape until the advent of myxomatosis in the 1950s. Aerial photographs taken shortly before the disease struck show large areas of open ground, heavily grazed by the rabbits and virtually clear of vegetation. Today the picture is dramatically different. Unchecked, the sea buckthorn that was originally planted before the First World War to stabilise the dunes has spread rapidly, and along with gorse and bracken has both choked out plants such as bee and pyramidal orchids and reduced the extent of suitable habitat for certain sun-loving invertebrates. Although the rabbits have since returned in limited numbers, proactive conservation work is required to open up the dune heath once more. This is done by scrub clearance, bracken spraying and the grazing of ponies and cattle to prevent future encroachment.

Clearly, rabbits can be a useful heathland management tool. At the Norfolk Wildlife Trust's Weeting Heath, specially managed for stone curlews, one of the objectives in fencing the site was to keep the rabbits *in*, to help ensure that the grazing pressure is sufficiently high to produce the very short turf required by the stone curlews for feeding and nesting. What this does mean, however, is that discrete colonies or stands of endangered heathland plants such as spiked speedwell must also be fenced, this time to keep the rabbits *out*!

Reptiles and amphibians

Lowland heath is of critical importance to Britain's modest reptilian and amphibian populations. It is the only habitat to support all six species of native reptile (grass snake,

Smooth snakes derive their name from the fact that their scales are entirely flat, and not with a central ridge (as with the grass snake and adder).

adder, smooth snake, slow-worm, sand lizard and common lizard), and is also significant for amphibians, particularly the natterjack toad. The conservation of heathland is therefore paramount to the protection of these species.

Arguably the most notable reptiles associated with heathland are also the rarest: the smooth snake and the sand lizard. Smooth snakes are a southern European species on the very edge of their range in Britain. Interestingly, on the Continent they are found in a wide range of habitats, but here they are restricted to southern English heaths and have only ever been recorded for certain in Berkshire, Dorset, Hampshire, Surrey, Sussex and Wiltshire. Formerly more widespread within this limited area, they have declined in the face of heathland destruction and deterioration and are now very localised (and totally extinct in Berkshire and Wiltshire). However, at certain sites they are not uncommon and although it is always difficult to be certain of their precise (or even approximate) numbers, there are signs that the majority of the surviving populations are at least holding their own.

Like all snakes, smooth snakes are shy and elusive, rapidly moving out of sight at the threat of disturbance. They prefer dry, sandy heaths, especially when on south-facing slopes, but will also use other adjacent habitat types such as grassland and woodland edge so long as these are not too shaded. Adults can reach a length of more than 60 centimetres (24 inches) and kill their prey – mostly small mammals and other reptiles such as common lizards and slow-worms – by constriction, rather than by the injection of venom. Females usually breed biennially and give birth to live young. However, the breeding performance of British smooth snakes is rather less successful than that of their Continental counterparts (i.e. less often, with fewer surviving young), doubtless as a result of living in a habitat that is less than ideal in terms of climate and available prey. In winter the snakes hibernate below ground level, digging themselves deep into the sand (often in rabbit burrows).

The sand lizard is the other classic heathland reptile. Whilst lowland heath is its main British habitat, it is by no means as restricted to this environment as the smooth snake. Populations of sand lizards can also be found in sand dunes, for example, and historically the species was much more extensive in distribution, ranging as far north as North Wales, and the Lancashire/Merseyside coast, where it still survives. However, during the twentieth century numbers declined sharply and many colonies died out completely. Only on the southern English heathlands – primarily in Dorset, Hampshire and Surrey – do significant populations persist.

As its name suggests, the critical factor for this lizard is sand: it needs adequate areas of fully exposed sand in which it can both bask and create nest-sites. Equally important, however, is sufficient adjacent cover into which it can run in times of danger. The requisite mosaic of habitats is most often found on slopes and banks on heathland, and the species tends to congregate in colonies where these conditions are best met. Unlike the more widespread common lizard, sand lizards do not give birth to live young but lay eggs in loose sandy soil and rely upon the sun to effect the hatching process, a risky strategy in the indifferent British climate. The adults are rather larger than their common relatives, reaching up to 20 centimetres ($7^1/_2$ inches) or so in total length from head to tail, and the males are considerably brighter in appearance, especially when in breeding condition, when their vivid green flanks are particularly noticeable.

During the mating season (May/June) the male sand lizard is an ostentatious creature, making the most of his colourful appearance. Fighting between competing males is commonplace, and females are often contested for by several potential partners.

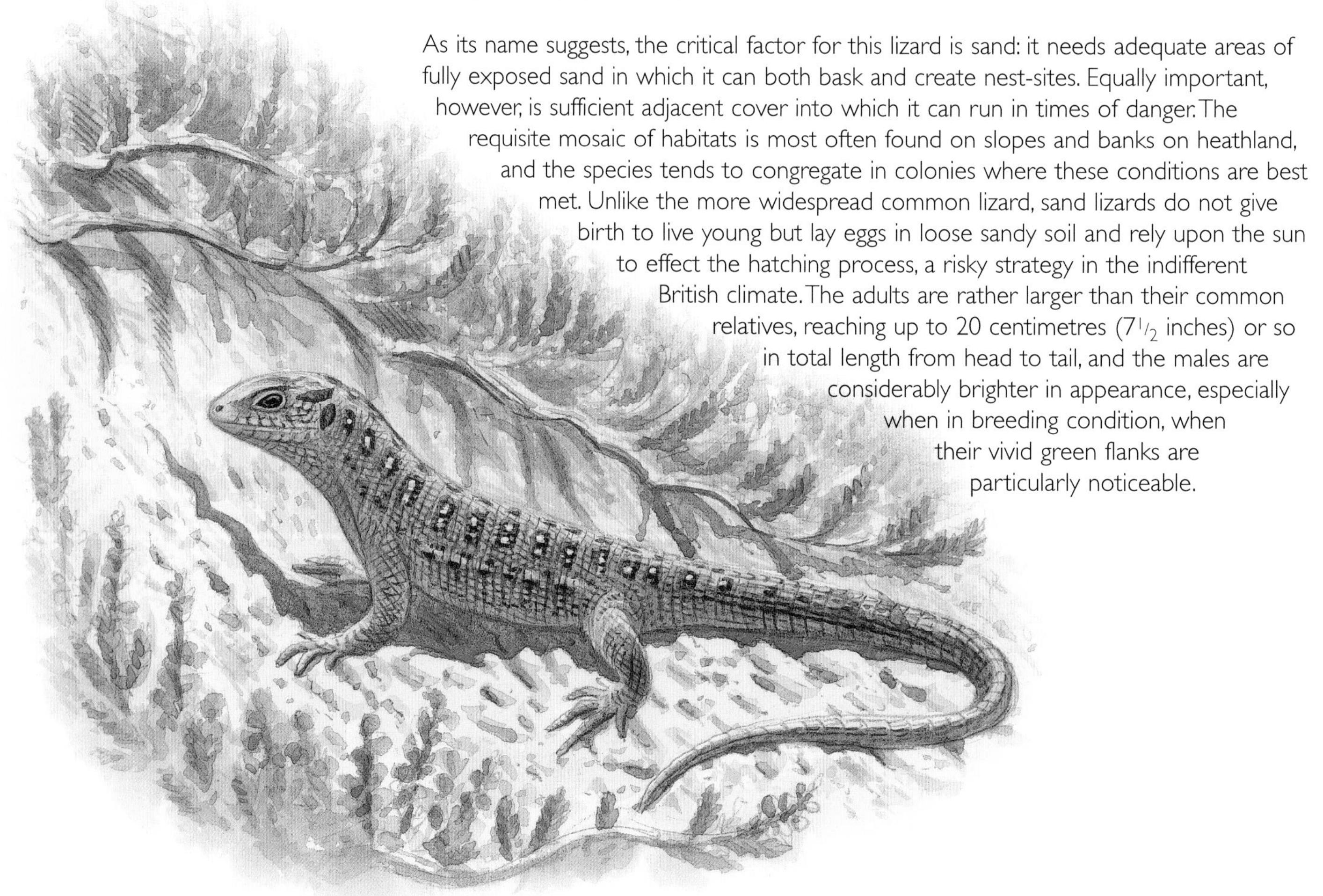

The sand lizard's threatened status in Britain has led to it being the focus of a Species Action Plan designed both to safeguard its habitat and manage this more effectively as a means of arresting the lizard's decline in its last remaining haunts. Meanwhile, as part of a Species Recovery Programme, sand lizards have been successfully reintroduced to suitable habitat in areas where they had previously died out – such as North Wales, Cornwall and Devon – as well as to other sites in counties where they had managed to survive but were in perilously low numbers and incapable of natural recolonisation, owing to habitat fragmentation. Although the future of the species is still not entirely secure, there are probably more sand lizards living wild in Britain now than for a century or so.

Heathland's star amphibian is undoubtedly the natterjack toad. Whilst common toads, frogs and the three British newt species will all use heathland habitats, the natterjack is more or less confined to heaths and to coastal marshes and dunes. It is dependent on the presence of shallow pools with low acidity and little aquatic vegetation, as well as on the availability of open sandy areas into which – rather like the smooth snake and sand lizard – it digs burrows. This is essential as a means of maintaining an equitable body temperature, and the nocturnal toad will often spend much of the day sitting in its burrow before emerging at nightfall in search of food. A gregarious creature which often forms sizeable colonies, the natterjack was formerly widely distributed in England and south-west Scotland, but is now very localised and scarce, with remnant populations left only in Cumbria, Lancashire, East Anglia and a few places in southern England. One of the main causes of its decline during the latter part of the twentieth century was the increasing unsuitability of much of its habitat due to the unchecked encroachment of vegetation, largely brought about by the demise of the rabbit population in the wake of myxomatosis in the 1950s.

Generally smaller than the common toad, the natterjack's shorter limbs make it unable to hop. However, it is a good runner and can move both faster and more nimbly than its commoner relative.

However, careful habitat management can successfully maintain and increase natterjack toad numbers, as at the National Trust's reserve at Formby near Liverpool, where artificial toad scrapes have played a central role in bolstering the local natterjack population (which lives on dunes rather than on heath). The scrapes are created from butyl liners and the water levels within them are managed to create conditions similar to the temporary storm-created pools that are the toads' natural breeding ground. With up to

50 toads seen at one pool in a single night, numbers are holding up well and natterjack spawn from Formby has been transferred to other dune sites to help secure the species's wider distribution.

Although reptiles and amphibians living on heaths are vulnerable to all the familiar threats to heathland wildlife, such as habitat destruction, fragmentation and disturbance, being largely sedentary and in some cases rather slow-moving they are particularly susceptible to the ravages of fire. They are often caught in the flames, and even those with the ability to burrow underground cannot always escape a 'deep burn', whereby a very hot fire will totally incinerate the upper litter layers. Even those that survive a fire are suddenly exposed to predators and soon picked off. The slow reproductive rate of species such as the smooth snake seriously inhibits repopulation following fires, and it can take several years for local population levels to climb back to pre-fire numbers.

Invertebrates

Lowland heath is of huge importance to invertebrates. Some 5,000 species are known to occur on heathland, a total boosted by the huge opportunities provided by the

Sandy tracks, as seen here at Frensham Common in Surrey, can be excellent habitat for sun-loving invertebrates, so long as the paths are not heavily used. Exposed areas of compacted sand and soil have generally declined on heaths with the demise of traditional human activities such as turf-cutting and mineral extraction.

variety of smaller habitats that exist within the heathland envelope. As heathland ecologist Nigel Webb points out, it is this mosaic that compensates for the relative lack of plant diversity in dwarf shrub communities, and for the fact that only a few invertebrates have a specific relationship with members of the heather family. However, many of the invertebrates present on heathland are generalists, equally prevalent in other habitats. It is more the specialists that are of particular interest. Overall, the range of heathland invertebrates is vast, from tiny mites operating in the litter below heather plants to large dragonflies wheeling over bog pools.

The microclimate of heathlands is of major importance to many invertebrates, in the sense that it offers a lifeline to those specialists requiring warm and dry conditions. Indeed, for certain sun- and heat-loving species the exposed, shade-free areas of sandy soil found on some heathlands provide their only viable habitat in Britain. Yet despite the huge number of different invertebrates potentially found on heathland, many are surprisingly vulnerable. For example, some species are very sedentary, incapable of relocating or of recolonising sites from which they have died out previously. Their requirements are often highly specific – for example, some species of burrowing and digging insect will only frequent open areas where the sand particles are of a certain size – and so the smallest change in local conditions can spell disaster.

The green tiger-beetle is a widespread and dramatic species, for which the drier types of heath provide very suitable conditions. Although predominantly at home on the ground, where it hunts down prey such as ants, it will fly readily when disturbed.

One of the most important groups of heathland invertebrates is also among the smallest in size: the ant family. Some twenty species are found on heathland, each occupying a particular mini-habitat. Along with vegetation density and height, soil type and moisture are critical to ant distribution, but most species (typically the aptly named turf ant) show a pronounced preference for the dry, sandy areas and patches of short grass in which they can easily construct their complex, communal nests. One exception is the black bog ant, which lives in amongst waterlogged vegetation in the valley mires of the New Forest, Isle of Purbeck and Gower Peninsula, for example. The presence of ants on heathland is critical to many other species of wildlife; they attract birds such as the green woodpecker, and are a vital part of the life cycle for certain types of butterfly (see below). They are also parasitised by several species of bee, fly and wasp.

Beetles are among the notable families of invertebrate associated with heathland, with hundreds of species occupying a range of specific niches across all types of heath, from dry to valley mire. Compacted bare ground and very thinly vegetated areas provide particularly good conditions for tiger-beetles, most characteristic of which is the green tiger-beetle. Armed with long legs for running down its prey (mainly ants) and large mandibles, this is one of heathland's most voracious invertebrate predators. It is fairly common on well-drained soils across much of Britain, but the tiger-beetle group also includes some particular rarities that are restricted to heathland. Foremost among these is the heath tiger-beetle, once widely distributed but now reduced to a handful of sites, mainly in Dorset. Other specialities include Kugelann's ground-beetle and the ground-beetle *Amara famelica*, now so endangered in this country that it is subject to

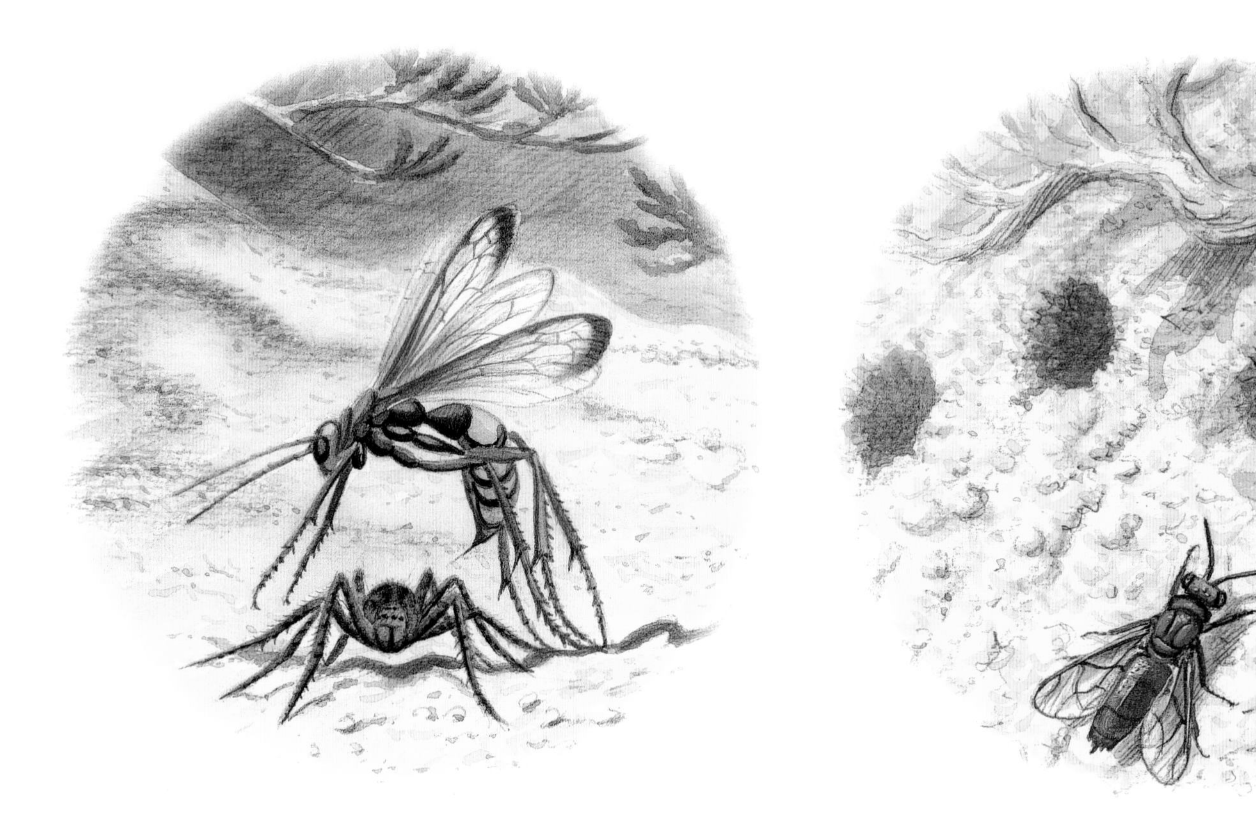

Sandy banks are ideal situations for invertebrates that can dig and burrow, and for those that parasitise such species, such as the black-banded spider wasp (above), and the ruby-tailed wasp (above right), seen here next to holes excavated by mining bees.

special efforts to bring it back from the brink of extinction. Once known from many sites across southern and eastern England, this species has been recorded from only two sites in the last thirty years; one of the British BAP targets is to reintroduce viable populations to five sites within its former range by 2010. Sandy patches are also important to beetles such as dor beetles and the minotaur, which is closely associated with rabbit dung. It digs a burrow in the sand, into which it rolls a piece of dung. It then lays an egg in the dung, on which the emerging larva feeds until it is ready to move above ground.

Other species of beetle depend on heathland vegetation rather than the physical conditions of heaths. These include leaf-beetles, the most significant of which is the heather beetle. This small, brown beetle is one of the few heathland invertebrates totally dependent on ling, which is required for all parts of its life cycle. In particular, its larvae feed on new heather shoots and when abundant can have a noticeable impact on areas of heather, stripping much of the new growth and even causing the death of plants. Increased levels of nitrogen in vegetation, usually prompted by atmospheric pollution or 'spray drift' from adjacent farmland, can boost heather beetle numbers and in extreme cases their impact on the heather can result in invasive species such as grasses becoming dominant on a localised basis (see also p.63). This usually occurs when an attack of heather beetle coincides with a period of drought. Although the beetle can have major implications in upland areas where grouse-rearing is important, it is rarely a cause for real concern on lowland heaths. Indeed, some authorities argue that heather beetle attack can actually assist heather by encouraging regeneration from lower stems or rootstock. Heather beetle numbers are naturally cyclical, however, and prone to great fluctuations; a period of abundance is invariably followed by a population collapse, often as a result of a corresponding increase in the parasitic wasp which lives off them.

The open sandy areas frequented by certain beetle species are also attractive to bees and wasps. Fifty per cent of British species have been recorded from certain heaths, most notably Chobham Common in Surrey. These include mining bees, digger wasps and bumble bees, the latter being especially important pollinators of heathland plants. Many have very specific requirements; the black-banded spider wasp, for example, specialises in hunting wolf spiders. The female wasp will alight on a bare patch of sand and excavate down with her legs, in search of spiders in their holes below. Upon finding one, the wasp will inject enough venom to paralyse it and then drag it to the surface. She will then dig a burrow, store the spider inside and lay her eggs upon it: the larvae then feed on the still alive but incapacitated spider. The digger wasp will operate in much the same way, although its choice of prey is more catholic; the larvae of moths and butterflies are especially popular.

Spiders are notable heathland invertebrates, and are found in all heathland habitats ranging from pine scrub through the litter under heather plants to areas of mire and bog. There are many different species, and the heaths of south-east Dorset, for example, are home to some two hundred species – a third or so of the total number recorded in

Heathland spiders can be both large, as in the case of the raft spider (below left), and colourful, as with the ladybird spider (below). The latter, once thought extinct in Britain, was rediscovered in 1979 in Dorset.

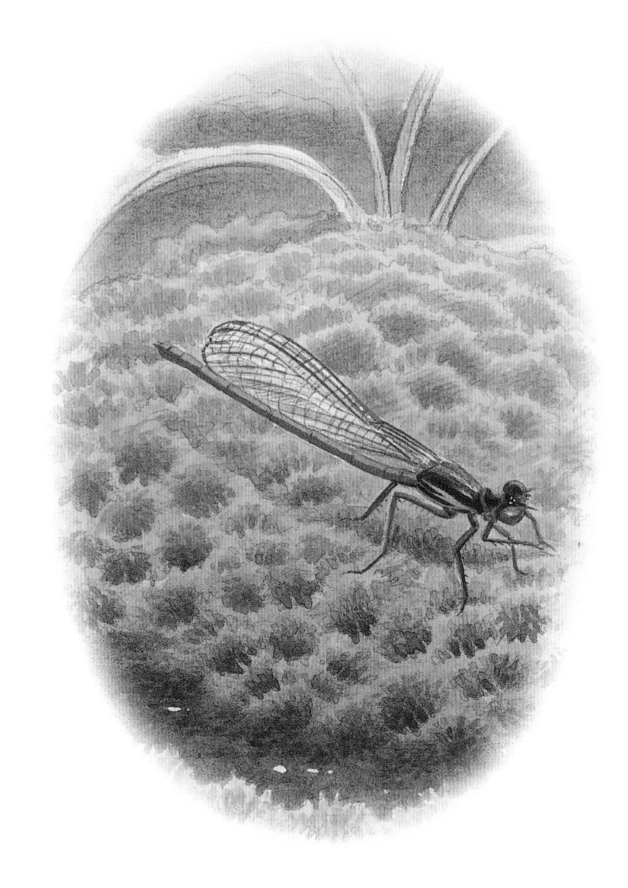

Heathland is noted for its dragon- and damselfly populations. The southern damselfly (above, trapped on a sundew) has very specific habitat requirements, and evidence suggests that numbers have declined at sites where grazing has ceased and where vigorous species such as purple moor-grass have been allowed to invade. The small red damselfly (above right, resting on *Sphagnum* moss) is also rare, and found mainly on heathland in southern England and west Wales.

Britain. Spider webs are certainly a frequent sight amongst heather and are usually of common species, such as the labyrinth spider, which spins a large sheet-like web up to 30 centimetres (12 inches) across. The very scarce and endangered ladybird spider is another heathland speciality; it preys particularly on tiger-beetles, which it ambushes from below from a short tunnel masked by a web. Although most spiders prefer dry habitats, one of the largest and most spectacular can literally walk on water: bog pools on southern English heathlands are home to the raft spider, which runs across the surface in pursuit of insects.

Areas of wet heath are also of great significance for both damselflies and dragonflies, particularly so when there are areas of open water, such as old peat diggings or gravel pits, or small streams running through mires. Many of the species recorded in Britain are at the northern edge of their range, and so have predominantly southerly distributions here. The heathlands of Dorset, Hampshire and Surrey are particularly good for them, and up to 15 species can be seen in one day at sites such as Studland on the Isle of Purbeck. Other, more common, varieties such as the four-spotted chaser can also be found in good numbers on heaths. Of the 52 species on the British list, 26 have been recorded at Thursley Common in Surrey. The small red damselfly and keeled skimmer are both associated with heathland in the sense that in this country they are more or less restricted to this habitat, but the real heath specialist is the globally threatened southern damselfly. This species has declined by some 30 per cent in the last forty years and is now found in fewer than sixty widely scattered sites, mostly in south-western Britain. It breeds along heathland streams and runnels and is under threat throughout its European range. All damsel- and dragonflies are sensitive to change, not only to the drainage and dredging of ponds and streams, but also to the vagaries of climate; for example, the demise of the rare white-faced darter at Thursley Common was attributed to a series of droughts.

Of the 60 or so species of butterfly recorded in Britain, relatively few are regularly present on lowland heath and only one can be claimed to have a specific association with this habitat. This is due largely to the low number of nectar sources on heathland; only in late summer, when the heathers are flowering, is there an abundant supply of nectar, but it is short-lived. Gorse may flower more generally, but it is not well favoured by butterflies. The low diversity of heathland plants is also a restrictive factor in terms of the choice of larval foodplants.

Although the names of three species of butterfly – large heath, small heath and heath fritillary – appear to signal an obvious link with heathland, the large heath is actually a butterfly of peat bogs, the small heath is widespread and common in a range of habitats from mountainsides to marshland and definitely not confined to heathland, and the heath fritillary, now very scarce, belongs more to woodland edges and clearings. Meanwhile, although familiar butterflies such as the red admiral, peacock, painted lady, meadow brown, common blue, small copper, large and small skippers and various members of the white family all occur on heaths from time to time, only a handful of species are real lowland heath specialists.

Perhaps the most notable heathland butterfly is the silver-studded blue. In Britain this species is virtually tied to heathland habitats, and its larval foodplants are all characteristic of lowland heath: ling, *Erica* spp., gorse, common bird's-foot trefoil, common rock-rose, horseshoe vetch and wild thyme. It is found on both dry and wet heaths, but like many British butterflies its population and distribution have shrunk dramatically in the last few decades, and it now occurs in just 20 per cent of its former range – and is still declining. However, in a few localities very large colonies continue to thrive, sometimes numbering tens of thousands. Such huge numbers can be seen on National Trust heathland on the western edge of the New Forest. The butterflies roost communally, and make for spectacular viewing in the early morning, when they can be watched *en masse* stretching their wings in preparation for their first flight of the day.

Although widely distributed, the silver-studded blue is a highly local and increasingly scarce species. Heathland is its most favoured habitat, although colonies are also found on coastal dunes and limestone outcrops.

One of the most fascinating features of the silver-studded blue is its close relationship with ants, which play a central role in the butterfly's life cycle. The butterfly lays its eggs in the colonies of either of two different species of ant, the ants subsequently tending the larvae and pupae, and even protecting the newly emerged adult butterfly before its wings are dry and it is able to fly. In return, the ants feed on the sticky secretions emitted by the larvae. This extraordinary relationship was discovered only a few decades ago and is still imperfectly understood.

The silver-studded blue is highly sedentary, and it is unusual for individual butterflies to move more than thirty metres (or yards) from their point of emergence during an entire lifespan of some four to five days. This explains why the species finds recolonisation so difficult, and why it is particularly vulnerable to habitat fragmentation and deterioration; when conditions in a particular locality become unfavourable (due, for example, to scrub encroachment), the butterflies are incapable of moving on and so simply die out. Even when conditions are good, their ability to take advantage of the opportunity is strictly limited. The rate of expansion at some North Wales colonies has been estimated to be no more than half a mile per decade!

The maritime heaths of south-west England were among the last British haunts of the large blue, which became extinct here in 1979. This species depended on a similarly symbiotic relationship with ants as its silver-studded relative, and was once plentiful on the heathland slopes of the Atlantic coast of Cornwall and Devon, where both the ants and its larval foodplant, wild thyme, were widespread. Much prized by collectors, it was ironically not so much their predations as myxomatosis that sealed the large blue's fate. Close grazing by rabbits helped maintain the very short sward required by the ants, but once the rabbits fell victim to the killer virus the turf became overgrown, the ants disappeared and the large blue was effectively doomed. However, a reintroduction programme (derived from large blue eggs brought from Sweden) has successfully restored the large blue to several sites in England. Two of these newly established populations are on National Trust heathland.

Both the green hairstreak and the grayling are frequent heathland inhabitants, but tend to favour heaths at rather different stages of development and growth. The grayling prefers open heaths with scant vegetation and extensive areas of bare soil, on which the immaculately camouflaged adults often rest. Males are strictly territorial and actively patrol their patch, with a distinctive looping and gliding flight. Some of the largest British colonies of grayling are on the heathlands of Dorset, Hampshire and Surrey, although elsewhere the species is predominantly coastal, an inhabitant of dune slacks. The green hairstreak, however, shows a marked preference for heaths with extensive areas of scrub, and commonly uses gorse as a larval foodplant. It is widely distributed across England, Wales, Ireland and Scotland, but is rather local and generally uncommon. Several species of fritillary can also occur on heathland or, more accurately, on heathland fringes where their required habitat and vegetation may be opportunistically – and temporarily – present. However, the much-declined marsh fritillary can thrive successfully on permanent wet heath and can be seen in these conditions on the Gower Peninsula in South Wales, for example.

Lowland heath is an important habitat for moths. Over thirty species use ling as a larval foodplant, and several of these use only ling. Locating moths in heathland is far easier than in most other habitats; simply walking across an area of heather will usually disturb a number of species, and many fly anyway during the day. These include two of the biggest and most widespread: the emperor and fox moths. The emperor is one of the most attractive British moths, the striking eye markings on its wings being characteristic. Interestingly, only the male flies by day; the female is strictly nocturnal. The fox moth is on the wing both by day and night, and in late summer its larvae are often a common sight on heather sprigs, where they feed up before hibernation. They overwinter as larvae before pupating the following spring and emerging as adults during May and June.

Other regular heathland moth species include the true lover's knot and aptly named common heath, both of which are widely distributed and at times abundant. The latter species flies regularly in sunshine, as does the silver-Y, an immigrant to Britain which often arrives in vast numbers and is attracted to the flowers of many different plant species. Heaths are also home to some notable moth rarities. These include species such as the speckled footman and shoulder-striped clover, which are both restricted to southern English heaths, and the marbled clover and tawny wave, both known primarily from the East Anglian Brecks. The conservation of these very localised species is now a matter of high priority.

A female emperor moth resting on ling, with two flying males attempting to locate her by picking up her pheromones (a chemical scent that stimulates male activity) with their antennae.

The male fox moth, shown here, is a rich brick colour, whereas the female is much greyer in tone. The impressive larvae are regularly predated by birds.

Two species of heathland grasshopper are also worthy of mention. Both are restricted to a few heaths in Dorset and the New Forest, but they favour different types of habitat. The heath grasshopper prefers dry locations, but the large marsh grasshopper is found only on wet bogs. It is at the most northerly extent of its European range and must be considered endangered. Also of note is the field-cricket, a flightless, ground-living species which is highly dependent on areas of dry, open grassland. Once found widely on grassy heathland, it is now found naturally at only one site in Britain, although reintroductions are taking place under an English Nature Species Recovery Programme.

The value of heathland to invertebrates has only recently been properly appreciated. Indeed, the precise nature and extent of its significance is still imperfectly understood in the case of some species, whose complete life cycles remain something of mystery. Until their requirements are better appreciated it may prove impossible to reverse declines in number or reductions in range. Indeed, in the case of some heaths the destruction or degradation of habitat has already caused the site extinction of important invertebrate species. In some instances this will have happened before we even knew what was there. More effective habitat protection is undoubtedly helping, but vulnerability to fire, the impact of certain (hopefully now redundant) management techniques, and a general inability on the part of many species to readily recolonise remain inhibiting factors. As if that were not enough, other factors may be beyond the immediate control of those caring for heathland; climate change, for example, may cause the desiccation of the heathland bogs and mires that are home to certain endangered species. Only time will tell.

First positively identified in Britain at Studland in 1933, the heath grasshopper is still known only from the heaths of Dorset and the New Forest.

Chapter Five

❖

Saving heathland

SAVING HEATHLAND

By the early 1980s it was clear that urgent action was required to save what was left of Britain's heathlands. Although many were already in safe hands, owned or managed by organisations such as the National Trust, Royal Society for the Protection of Birds, county wildlife trusts and English Nature (formerly the Nature Conservancy Council), others were placed rather more precariously. Some sites were still without any form of statutory protection, and meanwhile there were many owners who neither understood nor cared about the value of heathland and how best to manage it, or for whom the main priority was something other than spending time and resources looking after their heaths. Confusion over appropriate management techniques, and indeed over the need for management at all, compounded the situation.

Increased dialogue and greater awareness of the extent of the loss led to the start of real efforts to save heaths. In Britain, the 1981 Wildlife and Countryside Act proved successful at reinforcing the protection conferred to sites by SSSI designation, and the number of heathland sites so covered was increased accordingly (later legislation provided similar safeguards in Northern Ireland). Various subsequent European Union designations have further strengthened this protection. The launch in 1993 of English Nature's National Lowland Heathland Programme was also significant. This led to a national heathland strategy, which identified areas of surviving heath and included both recommendations on their management (for which grants were offered) and proposals for recovery programmes for scarce flora and fauna. Plans were also put in hand for restoration projects on sites where heathland had been destroyed or damaged.

Meanwhile, other landowners with significant heathland holdings, such as the Ministry of Defence and Forestry Commission, began to look more carefully at how best to manage heaths for conservation and reconcile this within the context of their own particular remits. However, the single most important development was the publication in 1994 of the United Kingdom Biodiversity Action Plan (UK BAP), born from the environmental concerns discussed by world leaders at the 1992 United Nations Conference on Environment and Development in Rio de Janeiro. The UK BAP identified fourteen key habitats for priority conservation, of which lowland heath was one. Targets were set of maintaining and restoring all existing areas of lowland heath (estimated at 58,000 hectares or 143,500 acres) and of re-creating a further 6,000 hectares (14,800 acres). The door was now open to serious government action and, more importantly, meaningful levels of funding.

Since then, impressive progress has been made. Lowland heaths across the United Kingdom are now in a healthier state than for decades, and probably busier than they have been for centuries: wardens, ecologists, biodiversity survey teams, access managers, conservation volunteers, scrub clearance working parties, as well as increasing numbers of grazing beasts, are all now out on the heaths. The Tomorrow's Heathland Heritage initiative (see p.133), launched in 1997, is a major driving force behind this presence. The scale of involvement is huge, but then so are the issues.

PREVIOUS PAGE The Agglestone on Studland Heath, with Studland Bay beyond. If it were not for the care and concern of local landowners, most notably the Bankes family, who gave much of the area to the National Trust in 1982, this outstanding tract of heathland could have met the same fate as that now covered by Bournemouth, across the bay.

Proactive management is required to maintain open heathland and prevent it from being overwhelmed, as here at Witley Common, where invasive bracken and birch scrub pose particular threats.

The need for management

Central to contemporary heathland conservation is the need to manage habitat. It was the decline and, in many places, almost total cessation of traditional practices such as grazing and burning that led to the loss of many heaths to scrub encroachment. Doing nothing is therefore not an option. However, with appropriate action supported by regular management the process of succession that all heaths undergo if left ungrazed can be arrested, reversed and then controlled. Once basic heathland conditions are in place, specific management can then be targeted at creating and maintaining the smaller habitat niches that are required by specialist species of flora and fauna. Continued monitoring and control of the growth and condition of those heathland plants that are desired, such as heather and gorse, is also necessary to avoid vegetation that is either overwhelmingly too mature or too young; the objective is usually to ensure the presence on each site of as great a mix of age classes as possible, thereby maximising the potential opportunities for wildlife with varying requirements. Heather and gorse management is therefore essential for a healthy heath, especially as the latter can invade the former and, when old, becomes leggy and open-structured, which reduces its value to wildlife. Equally vital is the control of unwelcome interlopers such as bracken.

The range of non-native invasive species is now even greater than ever, with the familiar problems caused by rhododendron now being equalled, at least, by the impact of newer arrivals. *Gaultheria shallon*, introduced to England as a garden shrub, is now a major problem on the heaths of Surrey and the New Forest, for example, and there is also growing concern over the spread of *Crassula helmsii* (Australian swamp stonecrop or New Zealand pigmy weed). This can form a dense mat over heathland pools and have a serious impact on species ranging from the natterjack toad and dragonflies to scarce plants such as pillwort. Proactive management will be required to deal with these problems. Nor does heathland management end with care of the habitat itself; it is also central to balancing the demands and impact of the public presence on heaths and to achieving a positive relationship between heathland and the wider landscape beyond, both of which are among the greatest challenges ahead.

Tree, scrub and bracken control

Left to its own devices, heathland turns into woodland. The process of successional change takes place as one type of vegetation is replaced by another, and on heathland this usually means the supplanting of the specialised heathland plant community (primarily heather and, to a lesser extent, gorse) by scrub and trees. Scots pine and birch are the most enthusiastic and rapid colonisers, but species such as hawthorn, blackthorn, oak, sycamore and, on wet heaths, sallow, will also invade readily. The result is the choking out of the more specialised plants. Active intervention is therefore required to physically remove the scrub and trees by cutting or pulling up seedlings and felling larger specimens, and then to introduce practices – such as grazing – that will maintain the heath's restored open aspect.

In recent years an enormous amount of time, energy and money has been (and still is being) invested in the clearance of scrub and tree cover from heathland. The transformation can be dramatic, and at some sites has been a source of disquiet and even opposition by local residents. Heathland enthusiasts apart, most people do not enjoy seeing trees felled, and accusations of environmental vandalism are never far away in such cases. Good public relations in advance of clearance is clearly key, but it is also worth noting that scrub and trees do have a role to play on heathland. They can constitute valuable habitat for birds, invertebrates and mammals, particularly so in terms of the 'edge effect' they provide, i.e. the interface between open heath and closed canopy, which can be specially valuable to moths, for example, and to birds such as the nightjar and tree pipit. Character trees can also enhance the heathland landscape – the twisted Scots pines of the Brecks and Surrey heaths are good examples – and also merit consideration as worthwhile components in the wider heathland context.

Bracken is perhaps less of an asset. Its legendary indestructible quality has led to endless debate and experimentation with potential 'final solutions'. Certainly total eradication is a tough proposition, but in most circumstances control is possible. Cutting does not kill bracken, as it usually reshoots; rolling – either with logs pulled by ponies, the traditional method, or by rollers towed behind a vehicle – may be more effective, as it crushes the bracken's cell structure and prevents the rhizome from replenishing the bruised fronds above. However, this must be done with care during the bird-nesting season and in areas where reptiles may be using the habitat. There is also the herbicide option, and one highly effective product – asulam – has been developed to specifically kill bracken but not seriously affect other plants and wildlife. Spraying by helicopter is sometimes used over large stands of bracken, serving as a symbol of the extraordinary lengths to which man now goes to remove a plant that was once a much-valued heathland product (see pp.67–8). Whatever method is used to control the bracken growth, it is essential then to remove the thick bed of dead bracken litter below, for it is this that prevents the regeneration of heather and other heathland plants.

The return of the beasts

Wherever possible, contemporary heathland management is centred around the re-establishment of traditional practices. This includes a return to grazing, arguably the single most important activity in creating and maintaining heaths. Livestock will inhibit scrub encroachment and invasive species, as well as help develop a mosaic of micro-habitats, which boosts wildlife diversity. However, the situation is one of considerable subtlety and finesse, as different breeds and types of animal affect the heathland environment in varying ways, their impact depending to a large extent on factors such as stocking density, vegetation type and structure, and duration of the grazing regime.

Most British heaths were ungrazed for at least half a century, from roughly the Second World War onwards. The only large heathland area where grazing has continued uninterrupted for hundreds of years is the New Forest. Only now is the practice being resumed on many other heaths, and not yet on all. Clearly, the world has moved on from 1945, and factors such as increased car ownership mean that today's livestock management is rather different from half a century ago. A more formal infrastructure is called for, ranging from extensive fencing and gridding to on-site interpretation and protection of the livestock against disturbance and wilful harm. Fencing is potentially controversial, and on common land can only be temporary (and even then requires the approval of the Secretary of State). Effective consultation and publicity of the value of grazing heathland are clearly very important. Experience shows, however, that the return of livestock to heaths is generally popular with local residents and visitors alike.

Not all heaths are suitable for grazing. The most appropriate are those with a mix of habitat types: wet and dry heath, with some scrub (useful as shelter) and open areas of grassland. Modern breeds of livestock are usually unsuitable for heathland grazing, their contemporary sensitivities rendering them unable to cope with the harsh environment. Traditional breeds are more hardy and often thrive, but all sites vary and the process at each is more one of experimentation than certainty. This also applies to stocking density, which is best operated as a 'rule of thumb' rather than being driven by prescription.

Different types of animal graze in different ways, and on different plants. Cattle, ponies and sheep are all effective at grazing deciduous scrub growth and grasses. Sheep and ponies tend to graze tightly to the ground, whereas cattle tear at the vegetation, leaving a more open sward. Ponies will deal effectively with gorse (especially when it is young) and will even eat bracken, which remains untouched by sheep and cattle. Sheep are at their most comfortable on drier heaths; ponies cope well with wetter areas. Goats also have value, as they will turn their attentions to woodier plants that other species may ignore. Their ability to handle life in testing terrain also makes them potentially suitable for use on clifftop heaths, for example. Pigs also have a role to play. Tamworths have been used on a heathland site in Dorset, for example, to control bracken; their foraging is so thorough that they uproot the bracken rhizomes, exposing them to frost. By

Welsh mountain pony browsing gorse on St David's Head, Pembrokeshire. Grazing animals can be both a practical management tool and a visitor attraction.

clearing the litter they also expose the latent heather seedbank below, encouraging germination. Nor is grazing the only benefit of a livestock presence on heathland. The churning action of hooves is very important for opening up areas within the vegetation, allowing less robust plant species to prosper, and it also helps distribute seeds around the site, thereby increasing diversity. There is an aesthetic element, too: animals grazing on a heath look good and create interest.

Demand for grazing animals on conservation sites such as heathland is now higher than it has been for decades, yet the number of suitable animals available has declined sharply in recent years. Changes in agriculture mean that many farmers have rid themselves of stock and are no longer interested in resuming their existing grazing rights over heaths and commons. Wherever possible, this is the preferred option, but many landowners have found it necessary to form their own flocks and herds and, if necessary, move them around from site to site: so-called 'flying flocks'. This has proved especially effective in Norfolk, where the local wildlife trust rotates its flock of sheep around the county, grazing them both on its own reserves and on other areas of heathland in which it is involved, and in Pembrokeshire, where the National Trust uses its own herd of Welsh black cattle to manage a network of heathland sites.

It may be heartening to see livestock wandering on our heaths once more, but grazing is not a cure-all. There are occasions when it may not be suitable or practicable, and in all cases the introduction of grazing beasts should be a planned and considered action rather than an automatic response. Equally, we should be under no illusions about the nature of contemporary grazing: the animals are there essentially as a management tool, and their role is far-removed from the economic role they played in the lives of past generations of heath-dwellers. Yet they remain a potent symbol of heathland's present-day renaissance.

Burning issues

For thousands of years the autumn and winter burning of heathland was a traditional practice. Fire removed old, dead and dying vegetation and encouraged fresh forage for livestock. On many heaths it ceased within relatively recent living memory, and on others has continued by default, owing to vandalism and arson. Burning both modifies the structure of heathland vegetation – useful if an age–class imbalance is developing – and also helps ensure that the nutrient level of the soil remains low, by burning off the litter that accumulates below the plants. It can also be a successful way of prompting the regeneration of heather in particular – the fire's heat stimulates germination from the latent seed-bank in the soil – and of preventing the ascendancy of more invasive species such as purple moor-grass. Burning is therefore a valued management tool on certain heaths, but one that is practised only with great care and on a highly controlled basis.

In England, strict guidelines on burning are issued by the Department for Environment, Food and Rural Affairs (DEFRA), and it is illegal between 31 March and 1 November because of concern over the potential harm to wildlife. When a burn is planned, the area is carefully prepared, with a firebreak cut around it and water supplies on hand to extinguish the flames once the desired area is burnt. Specialist advice is taken on wind direction and strength, and the area checked for sedentary wildlife species that may not be able to flee the flames in time. Even with such procedures in place, burning is not appropriate on all heaths. There are concerns, for example, about its impact on lichen and moss communities, and on reptile and invertebrate populations particularly, and as heather can take up to two years to re-establish following a fire there can be a problem with opportunistic species muscling in before it gets going. In such cases, other methods of heather management include mowing using machinery such as a forage harvester, which cuts the heather to a height of two to three centimetres (one inch) by means of a rotating cylinder with protruding blades. This stimulates the heather to regenerate from its rootstock, and in today's age of heathland restoration also has the distinct advantage of providing heather cuttings – known as 'brash' – for use as stock material for heathland re-creation projects (see below).

Open to visitors

The issue of public access on heathland is complex. Heaths have always been associated with human beings, and are directly affected by the way in which we utilise and manage them. Indeed, one of the main problems with heathland in recent decades has been the lack of human management. Public access to heaths is often enshrined in common rights, and a continued human presence there is certainly desirable. Heathland can offer both valuable green space and excellent educational and recreational opportunities, and is increasingly a showcase for contemporary conservation and landscape management. Visitors are therefore welcome on most heaths, and are arriving in ever-increasing numbers; the National Trust's Studland Heath in Dorset, for example, receives 1.5 million visitors each year. Meanwhile, on-site facilities are growing, with more purpose-built visitor centres, increased interpretation and expanding special events programmes.

However, if not managed carefully, public access can have a serious impact, particularly at the more accessible and popular sites. In the vast majority of cases visitors will keep to footpaths, so waymarked routes are an effective method of presenting a site's main points of interest whilst also keeping disturbance of more sensitive wildlife species and habitats to a minimum. However, in areas of heavy access, soil erosion and exposure to damaging water run-off can result, and access over bare areas – of huge importance to invertebrates – can be disastrous if uncontrolled. Other types of access, such as horse riding, can result in the trampling of vegetation and the creation of new paths through heather, and both motor-biking and mountain biking can cause serious disturbance and soil erosion. In such cases, the policy of most landowners is to permit such access on a controlled basis wherever possible, including the provision of discrete areas for these activities. The installation of restrictive access points can help reduce the impact elsewhere on site.

Dog-walking is an especially sensitive issue. The use of open space, including heathland, for exercising dogs has almost assumed the status of a new common right. Yet there are serious concerns about dogs on heaths, mostly arising from the impact of uncontrolled animals on ground-nesting birds, reptiles and – of increasing significance with the return

Heather will usually regenerate rapidly after fires, the heat helping to stimulate the germination of the dormant seed-bank. Meanwhile, the bare areas of soil exposed by fires are attractive to birds such as woodlarks, which will often move onto a heath shortly after it has been burned.

When exposed, sandy heathland soils are highly prone to erosion and soon break down under excessive weight of feet (both human and equine), especially when coinciding with periods of heavy rainfall. Such pressure can lead to localised scarring, as here at Upton Heath in Dorset.

Located in the heart of the Surrey heathlands, the National Trust's Witley Centre receives many thousands of visitors each year and is a focal point for visits to the area.

of grazing to so many heaths – livestock. The control of dogs at all times is therefore a prerequisite. However, the dog issue is not restricted to disturbance to wildlife and livestock; through their urine and excreta, dogs can enrich heathland soils, and although this is usually limited to path verges, it can have an impact on sensitive plants dependent on low levels of nutrients. At the National Trust's Headley Common in Surrey, for example, research revealed that 1.5 tonnes of dog excreta were being deposited on the heath every year, much of it within a relatively small radius of the car park. In an attempt to reduce this figure, dog bins were provided and dog-owners encouraged to clear up after their pets. This appears to have had some modest success, not least in terms of raising awareness of the problem.

Much has been made recently of the implications of the 'right to roam' provision within the Countryside and Rights of Way Act (2000). Many areas of heathland already have *de facto* open access, by virtue of tradition, formal right or permission of the owner, and so the impact of the new right of access enshrined in the Act is unlikely to be as far-reaching as first expected. What the Act does emphasise, however, is the need for visitors to behave responsibly, and for landowners and managers to offer as great a degree of access as possible, in return for the right to restrict access, often on a seasonal basis only, to those areas that are particularly sensitive in terms of wildlife conservation, for example. Through careful management, visitor access to heathland can be a rewarding experience all round, and one that marks the return of the people to the heath after decades of decline and neglect.

Urban heaths

The fragmentation of heaths, and the fact that so many have been built on, have led to some now being on the urban fringe, directly abutting residential or commercial development. Others are almost entirely surrounded by housing and by other potentially unsympathetic neighbours, such as industry, and so completely divorced from their original rural context. Not surprisingly, many such heaths are subject to intense public access, which has an impact, especially on ground-nesting birds and reptiles (both groups are highly vulnerable to dogs and cats, for example), and recreational activities such as motorbike scrambling are more frequent on urban heaths than elsewhere. Natural water levels can be affected by water control methods such as drainage channels, and by pollutants seeping from industrial plants or from sewerage system discharges. For example, two streams which drain urban heaths in the Poole–Bournemouth conurbation suffer regularly from background pollution of this type.

The installation of service infrastructures – roads, water and gas pipes, telephone and power cables etc. – can also be disruptive on urban heaths. Roads in particular can have a serious impact, acting as conduits for pollution, barriers to highly sedentary wildlife species and hazards to the more mobile. For example, research at Holt Heath in Dorset has revealed collision with road vehicles to be an important contributor to Dartford warbler mortality. Vandalism and fire are also major concerns on urban heaths. The tipping of rubbish and wanton destruction of 'conservation infrastructure' (interpretation panels, nest boxes, boardwalks, etc.) are enduring issues, and fire remains one of the single biggest problems. The contemporary fashion for torching joy-ridden cars accounts for many fires, but arson has always been a problem on heathland. Small fragmented heaths, such as those typical of urban areas, can lose a disproportionately high percentage of their vegetation cover to fire, especially when repeatedly burned. And urban heaths are burned much more often than their rural counterparts. Research by Lesley Haskins in Dorset highlights the problem: urban Canford Heath suffered 179 fires between 1990 and 2000, whereas rural Hartland Moor recorded just two, with data from elsewhere supporting this pattern.

The dumping and torching of stolen vehicles is an increasing problem on many heaths, especially those on the urban fringe. Not only does this look unsightly and raise security concerns in the minds of potential visitors, but it can also trigger devastating fires.

Firemen extinguishing a heathland blaze. Whether by accident or design, many hectares of heath succumb to 'unofficial' fires every year. For wildlife on small, fragmented sites the impact can be disastrous.

Yet despite this barrage of problems, urban heaths can be highly successful habitats, both in terms of their wildlife value and in what they can offer local residents as amenity space and a 'green lung'. Investment costs may be higher: there is often a greater need for infrastructure such as firebreaks, fencing, controlled access points and warden presence than on rural heathland, but many of the problems can be overcome with the support of local residents; once they value their local heath, many of the more anti-social and damaging activities that take place on it decrease or disappear completely.

Post-industrial heathlands

Economic changes during the late twentieth century freed up land that for several decades previously had been locked into intensive industrial use. As heavy industries such as coal-mining declined and, in some cases, expired totally, many hectares of derelict land became available. Much of this land was seriously affected by industrial occupation, its landscape transfigured beyond obvious recognition by features such as spoil-heaps and slurry lagoons, with a high incidence also of waste contamination and hazardous post-industrial infrastructure, such as shafts and pits. Proposals on how to deal with the complex issues involved rarely went beyond cosmetic – and often inappropriate – forms of landscaping, or further than simply excluding the public and leaving the site.

However, the industrial cycle can actually create a range of excellent opportunities for wildlife and, ultimately, for landscape enhancement. In many instances industries were established on heathland sites, and their demise offers scope for restoration and re-creation. Many plant and animal species need no invitation to colonise, even in the most unpromising circumstances, and natural revegetation can be surprisingly rapid. For example, at Stoneyhill near Telford in Shropshire, open-cast coal- and clay-mining (which ceased in the mid-1960s) left a series of pools surrounded by mounds of coal slag and clay spoil, often of widely differing acidity levels. A valuable and highly diverse mosaic of heath, grassland, scrub and open water soon established naturally, and there are similar examples from pit-mounds elsewhere in the Telford area. As ecologist John Box has pointed out, the wide range of plant communities that is typical of these sites is partly a result of the long history of industrialisation in the region; nature has been recolonising ever since the abandonment of some workings as long ago as the late nineteenth century, and so there is a range of 'opportunity niches' going back a hundred years and more.

National Trust wardens spreading heather brash at Heysham Head in Lancashire. The brash was collected from another Trust property, Thurstaston Common on the Wirral, and used to help regenerate an area of heath that had once been covered in tarmac and used as a go-karting track.

In some circumstances, however, nature may need a helping hand to reclaim the land rather more speedily than might otherwise be the case. At Rufford Colliery near Mansfield in Nottinghamshire, a heathland has been successfully established on a spoil-heap by a partnership led by UK Coal, Nottinghamshire Wildlife Trust and the county council. The surface of the highly compacted heap was initially 'ripped' or incised to prevent the development of a perched water table, then covered with locally derived sand to provide a suitable substrate for restoration; it was then sparsely sown with an acid grass mix, to provide a nurse sward for the heather, which was introduced in 1997 by spreading brash, i.e. stalks complete with seed-heads, gathered from local heathland sites. Establishment was surprisingly rapid, with the heather germinating in its first year, and an excellent heathland/acid grassland community has resulted. A slurry lagoon on the crown of the tip has been retained as a wetland area, adding further diversity.

Meanwhile, the Tomorrow's Heathland Heritage programme (see below) is funding the re-creation of 750 hectares (1,850 acres) of heath on clay spoil-tips in Cornwall, with English Nature and the two principal clay extraction companies as partners. Such ventures should help build conservation criteria into the planning of future industrial sites. Managing post-industrial heathland may yet prove to be a greater challenge than

Careful habitat management can help boost nightjar numbers on heathland. The birds breed and roost on the bare ground exposed by gaps in the heather canopy, and will take readily to areas created artificially by the selective uprooting of heather plants.

anticipated, however, simply because there is not much track history to go on. What is clear, however, is that landscapes of this type can form valuable heathland habitat and are also of great amenity value.

Management for important wildlife species

As heathland supports so many rare and often highly localised species, several of them of major conservation importance, considerable management resources are devoted to trying to meet their needs. The general approach is habitat-driven, i.e. if all the pieces of the habitat mosaic are in place, then in theory not much extra effort is required to create the conditions on which individual species of wildlife depend. However, for creatures such as the sand lizard, which has highly specific requirements, particular forms of management can help maintain and increase populations. The lizard needs unshaded, bare areas of sand in which to lay its eggs, and these can be created by turf stripping, either by hand or mechanically, and by then ensuring that they do not become invaded by vegetation. Maintenance effort is therefore high, but manmade areas such as this have been highly successful on the Purbeck heaths in Dorset. Equally, nightjars can be encouraged by the judicious thinning of encroaching scrub and woodland; they require open glades in which to hunt, and bare areas of ground on which to nest. But remove all the trees, and the nightjars will often leave also, as they require elevated song posts from which to deliver their characteristic churring.

Effort also goes into creating the appropriate conditions for species which may only use heathland at certain times of year, for particular purposes. For example, the chough requires areas of open turf during the summer, when it feeds its young on larvae dug out from the 'lawns' that are a feature of the grassland/maritime heath mix on the Pembrokeshire and Cornish coasts in particular. The National Trust is working with local farmers and other owners to ensure that adequate lawns are maintained and new ones created, thereby helping to boost the chough's scattered and vulnerable populations.

From farmland back to heath

The loss of so much heathland to agricultural use through the centuries makes it highly appropriate that resources are now being invested into reversing that trend. At Hartland Moor in Dorset, the National Trust and English Nature have worked closely over the last six years to restore 162 hectares (400 acres) of farmland to lowland heath. The two farms concerned were ploughed from the moor between 1950 and 1979, but by 1990 were no longer commercially viable and farming ceased in 1993. The decision was then taken to return the land to heath, although this had not been attempted on such a large scale in Britain before.

One of the key problems in returning agricultural land to heath is the build-up in the soil during the farming years of nutrients, especially phosphates, which inhibit the regrowth of heather in particular. This can be countered by the application of nitrogen fertiliser to encourage grass growth, which is then regularly cut; this reduces the nutrient levels to the point at which heather can begin to thrive. The planting and cropping of barley is also successful at drawing the nutrients from the soil. This then allows heather to

recolonise, and although in some areas at Hartland it is regenerating naturally, in others it proved necessary to spread seed gathered from neighbouring areas of heathland. The overall objective is to encourage the development of a mixed grass/heather heath, although the more vigorous heather risks overwhelming the grasses if left unchecked. The restored areas of heath are currently being grazed by cattle, although in future it is hoped that sheep-grazing may help achieve the required balance. Whatever regime proves successful, this is a very long-term project: it may take decades to produce the desired effect, and yet it is exactly large-scale, ambitious projects such as this that will make a real difference in terms of our national heathland landscape.

An ambitious scheme on Hartland Moor in Dorset will see the shaded area returned to heath within the next few years. Work is already underway and the heathland vegetation is recolonising naturally, albeit slowly.

Unexpected allies

Heathland conservation is happening on a larger scale than ever before, and in some unlikely quarters. Some of the most significant heaths are owned by organisations whose main responsibilities lay primarily in directions other than landscape conservation, some of which seem intrinsically incompatible with heathland preservation. But there has been a sea change in recent years and surprising alliances have been forged. Nowhere is this more apparent than with the Forestry Commission, whose business was always one of trees and timber production. Many of its plantations were established on tracts of heath: in Nottinghamshire, for example, there are 5,500 hectares (13,600 acres) of forestry, of which 4,600 hectares (11,400 acres) are planted on heaths. The notion that heathland conservation should therefore be a priority seems a paradox, yet that is precisely what is happening through the Commission's estates management agency, Forest Enterprise. Conservation awareness on forestry land is nothing new, however, and the management of plantation rides for wildlife interest is well established; rides occupy fifteen per cent of forest area on average and can be rich in specialist heathland plants and invertebrates. Most are mown or forage harvested annually to ensure continued diversity and control of invasive species. There has also been a policy of developing open areas within plantations, not least for landscape value. However, conservation objectives are now far more ambitious and directly related to Forest Enterprise's growing role as an amenity and recreational provider, within which conservation is increasingly important and, arguably, timber production less so.

Ephemeral heathland is a fact of life in many plantations, where the heather seedbank remains buried under the shadow of the trees, awaiting its chance. This comes when the trees mature, are felled and the ground litter cleared for replanting; the heather regenerates, only to be smothered after a few years by the young trees. However, this ability to rise from the ashes clearly makes for ideal restoration opportunities if the decision is taken not to replant. In Thetford Forest on the Norfolk/Suffolk border, 300 hectares (740 acres) of former forestry land is being restored to a mixed grass/heather heath, relying on the dormant seedbank in the soil rather than on the need to import seed from off-site. The resulting heath will then be grazed. The indications are that it might be as expensive to 'grow' and maintain heathland as it is to grow trees, but in an age of depressed timber prices and of increased demand for recreational landscapes that also offer wildlife value, this is perhaps the most appropriate option.

The Ministry of Defence is currently the single biggest owner of heathland in Britain and responsible for the care and management of 32 per cent of the UK lowland heath resource, at 53 widely scattered sites across England, Wales, Scotland and Northern Ireland. As discussed earlier (see p.54), much of this land came to the military authorities during the first half of the twentieth century, a time when heathland was under increasing pressure. Much of it has survived and even prospered under military management, to the extent that MoD-managed heaths are now among the richest of their type in terms of wildlife interest. Although the MoD understandably manages its land primarily for military use, wildlife conservation has been accorded a high priority for more than twenty years and is considered alongside other important issues, such as agriculture, water management, cultural heritage and public access, in site-specific management plans.

This integrated approach is readily apparent on the Stanford Training Area, which covers some 8,700 hectares (21,500 acres) of the East Anglian Brecks. The bulk of the military presence here dates from the Second World War, and STANTA is now one of Europe's busiest military training areas, used by some 80,000–100,000 troops annually. This pressure notwithstanding, the area is of considerable wildlife value and scenically unique in the sense that it includes the largest surviving remnant of traditional Breckland heath. The varied terrain, which is what makes STANTA so valuable for military training purposes, also provides a mosaic of heathland habitats on which a diversity of wildlife depends,

Heather growing alongside rides on Forestry Commission land at Clipstone Forest, Nottinghamshire. Heather has readily regenerated on some cleared areas of the forest, despite having been under plantations for six decades.

much of it scarce and local elsewhere. Conservation work is proactive, in the sense that military operations are, wherever possible, deliberately avoided close to sites of particular significance to wildlife. Equally, heathland restoration targets have been established and even exceeded, at some 320 hectares (790 acres) to date, and special breeding plots for stone curlews have been created and readily adopted by the nesting birds.

Tomorrow's Heathland Heritage (THH)

By far the most significant heathland initiative ever in the United Kingdom, THH was created as a result of the UK BAP and launched in 1997. Driven by English Nature, the lead agency on the UK BAP Group, the objective of THH is to work towards achieving over 70 per cent of the BAP 58,000-hectare (143,500-acre) heathland restoration target and 40 per cent of the re-creation target of 6,000 hectares (14,800 acres). The delivery of public access is also an essential component of the programme, which is scheduled to end in 2008 and is supported by the Heritage Lottery Fund to the tune of £14 million. The consortium of partners – over 145 in total, ranging from local businesses and community groups to large land-owning organisations such as the National Trust, Royal Society for the Protection of Birds and Forest Enterprise – has pledged an additional £12 million, making the total investment in heathland not far short of £30 million.

The marsh gentian is a localised plant and highly sensitive to encroachment from invasive species such as purple moor-grass. Several Tomorrow's Heathland Heritage projects are focusing on controlling the latter so that gentian colonies can expand.

As well as funding practical heathland restoration projects, the Tomorrow's Heathland Heritage initiative has also helped encourage greater understanding of the social history of heathland. As part of the 'Hardy's Egdon Heath' project, a short play entitled *Heathland Wisdom* – which looked at the insight of traditional heathland folk – was commissioned and performed to audiences across Dorset.

There are currently 25 THH projects, from Northern Ireland and Orkney south to Cornwall, containing a range of individual initiatives at multiple sites. The original expectations were of only nine or ten projects, but the response was so favourable that the scope of the programme was extended accordingly. A broad definition of heathland was taken, based more on landscape and cultural criteria than strictly ecological factors, with funding available for set-up costs and capital investment, which in some cases has included land purchase. 'Hardware' such as fencing, machinery for scrub clearance and bracken control, interpretation materials, etc., are all eligible for funding, and a large number of project-related posts have been created to implement the restoration and re-creation schemes. These are very varied, from the re-establishment of heathland on china clay-mining spoil-heaps in Cornwall to the clearance of willow scrub on the wet heaths around St David's airfield in Pembrokeshire, bracken spraying and harvesting (with a view to composting) on Cannock Chase in Staffordshire, the control of invasive gorse at Bloody Bridge in County Down and the linking of heathland fragments in Hardy's Dorset.

The project is already vaunted as a suitable model for the conservation of other types of landscape. The acid test will come when the set-up funds are withdrawn at the end of each project's life. Thereafter it will be the responsibility of the partners concerned to maintain the restored and re-created heaths from their own resources.

A heathland renaissance – what next?

After more than two centuries of decline and demise, Britain's heaths are on their way back. Many are already being restored and are in a healthier condition than for decades;

the next few years will see the renewal of many more, brought back from the brink of extinction by a combination of lottery funds and a return to traditional forms of heathland management. The amount of 'active' heathland should be virtually doubled under current initiatives, restoring the profile of one of our most significant cultural landscapes. Certainly cause for celebration.

However, we must also remember exactly what we are bringing back. These remanaged heaths will provide fine wildlife habitat, excellent recreational and amenity space, and a superb educational resource. Yet our newly restored heaths will not be truly working landscapes. Grazed yes, burned, yes (sometimes), but not worked in the way that throughout history was so defining a factor in their content and appearance, and which gave them such huge cultural value. The age of the furze-gatherer may have gone, a victim of economic change as much as anything, but with it ended a cultural continuity going back thousands of years. People may be back on the heath again, and in some numbers, but this is a different sort of interaction: one of recreation, and the presence of dog-walkers, birdwatchers or horse-riders can never be a substitute for the contribution made by those who toiled over the gorse, turf and bracken, trying to make a living.

Understanding this aspect of heathland is not straightforward. In the rush to draw up management plans, secure the future of BAP species and eradicate bracken, we may have lost sight of what heathlands are really all about. Certainly, much of their direct social evidence has gone with the passing of the last practitioners of certain crafts and techniques. We may not be able to restore this element as we can a tract of heather, but we can do more to explore and explain heathland's cultural significance.

Arthur Nash of Tadley in Hampshire is one of Britain's few remaining broom-makers. The future for one of the last surviving heathland crafts looks increasingly bleak, undercut by cheaper – and generally inferior – imports from overseas.

The challenge is to think big. Central to the ethos of initiatives such as Tomorrow's Heathland Heritage is the concept of linking heath fragments, of 'filling in the gaps'. This is precisely what we must do with heathland's social history: think big, look at the wider canvas and make these connections. The European Union-funded Heathcult project is currently attempting to do precisely this across seven countries in north-west Europe, including Britain, by focusing on the cultural significance of heathland and documenting the complex socio-economic structure that underlay European heathlands until very recently in some countries. It is this sort of understanding of heathland's cultural and social contribution that we need to get under way in Britain, to complement the huge advances made in heathland management and wildlife conservation. We should also be working towards a National Heathland Centre, along the lines of the excellent Lyngheisenteret in Norway. Only then shall we achieve a truly integrated approach to heathland conservation, and one that pays due recognition to the enormous contribution made to our heritage by our heaths, and by those who worked them.

The acrobatic hobby is one of Britain's fastest birds of prey and an increasingly common sight over our heathlands.

FACING PAGE The Surrey heaths are noted for their dramatic stands of Scots pine, as here at Frensham Common.

Places to visit

The National Trust cares for almost 250,000 hectares (617,750 acres) of the most beautiful countryside in England, Wales and Northern Ireland. Of this total, some 4,500 hectares (11,120 acres) are lowland heath, representing an important percentage of Britain's surviving heathland. Proactive management is designed to ensure that this valuable habitat is both maintained and enhanced, thereby benefiting the specialised wildlife that depends upon it.

The places to visit that follow are just a selection; fuller information on many more National Trust properties of landscape and wildlife interest is available in the Trust's *Coast & Countryside Handbook* or from the network of Trust regional offices, details of which can be obtained from the following address:

The National Trust Membership Department, PO Box 39, Bromley, Kent, BR1 3XL
tel. 0870 458 4000, fax 020 8466 6824, enquiries@thenationaltrust.org.uk

The Lizard Peninsula in Cornwall, an outstanding area of maritime heath and home to a range of highly local and endangered plants.

County Dorset

Property Studland Heath National Nature Reserve

Studland Heath and adjacent Godlingston Heath together form one of the most important tracts of lowland heath remaining in Britain. Dominated by the immense Agglestone, from which there are sweeping views, the heathland is home to a range of rare and specialised wildlife. This includes healthy populations of birds such as Dartford warbler, nightjar and stonechat, as well as good numbers of reptiles, notably sand lizard and smooth snake, although these are not easily seen. There are also many interesting plants, including the local Dorset heath and, in wetter areas, sundew and marsh gentian. Invertebrates include silver-studded blue butterfly and such rarities as the raft spider.

Access

Immediately north of Studland village, off the minor road leading to Shell Bay (and thence to Sandbanks by ferry). Open access to most areas of heath throughout the year via a network of footpaths and bridleways; four car parks (charge for non-National Trust members), with visitor centre, shop and café at Knoll beach. Two waymarked nature trails, programme of guided walks and study base for educational visits. There are two bird hides overlooking the Little Sea, an inland lagoon. Dogs must be kept under control at all times and during July and August are not permitted on some of the beaches adjoining the heath. Further information is available on 01929 450259.

When to visit

The second half of August is best for the remarkable show of four species of heather, all in bloom simultaneously. Still and warm evenings in June and July (from about 10pm onwards) are ideal for hearing the enigmatic song of the nightjar and, possibly, glimpsing the birds as they wheel around in the twilight.

County Suffolk

Property Dunwich Heath

An important remnant of the once extensive – but now highly fragmented – Suffolk coastal heaths known as the Sandlings, Dunwich Heath was traditionally maintained by extensive sheep-grazing. It is now a major site for wildlife, supporting many species of threatened flora and fauna. These include breeding Dartford warblers and nightjars, a winter hen harrier roost, and a range of interesting invertebrates, most notably the rare and peculiar ant-lion. Conservation management, such as heather-cutting and scrub clearance, helps ensure a diversity of open heathland habitats, ranging from bare ground and pioneer heathland to woodland edge.

Access

One mile south of Dunwich village, signposted from the A12. Open access to the heath via footpaths. Car park (charge for non-National Trust members), with information point, shop and licensed tea-room in the Coastguard Cottages. Education base and events, including guided walks. Further details are available on 01728 648501.

When to visit

Autumn is a particularly beautiful time at Dunwich. The heather and gorse are both in bloom until October, and many migratory birds can be observed passing through at this time of year. The car park gives fine views out to sea and over the adjacent and internationally renowned Minsmere reserve, managed by the Royal Society for the Protection of Birds.

County Surrey

Property Hindhead and the Devil's Punch Bowl

The commons around Hindhead were a battleground for early conservationists, who strove to protect the area from enclosure and development. Indeed, an important chapter in the early history of the National Trust was played out on these heaths (see p.34). Today they represent one of the most extensive areas of lowland heath in southern England. The Punch Bowl is a natural amphitheatre, its slopes cloaked in heather and with areas of scrub and woodland. Bisected by small streams, it was created by springs cutting through the soft underlying rock, and is the largest spring-formed feature in Britain. A range of heathland birds is found here, including hobby, nightjar, stonechat and woodlark, as well as interesting plants such as bog asphodel, bog bean and sundew. Exmoor ponies and Highland cattle are used to control the spread of birch, pine, bracken and coarse grasses, although some stands of woodland are retained as they constitute valuable habitat for birds such as woodpeckers and warblers.

Access

There is a National Trust car park (charge for non-Trust members), with café and information point, just east of the junction of the A3 and A287, north west of Haslemere. The whole area can be explored via a network of footpaths and bridleways. The Trust's Witley Centre, a few miles to the north east off the A3/A283, has information about the Surrey heaths in general. Further details are available on 01428 604040 (Hindhead) and 01428 683207 (Witley Centre).

When to visit

Hindhead is interesting throughout the year, but can be particularly stunning in July/August and during the winter months. The latter may seem an unusual time of year to explore a heath, but in winter Hindhead is the haunt of scarce birds such as great grey shrike, hen harrier and merlin, and its desolate yet evocative aspect captures perfectly the sense of awe and foreboding felt by early travellers to the area.

County Pembrokeshire

Property St David's Head

One of the most dramatic headlands in Wales, St David's Head is also one of the largest expanses of coastal heathland in Britain. The exposed seaward slopes are covered in classic maritime heath, the gorse and heather cropped short by the effects of salt spray and wind. Further inland, the heath is more diverse botanically, with heath bedstraw, heath milkwort, heath spotted orchid and lousewort all present, along with a variety of grasses and sedges in the wetter areas. Ponies graze the heathland, and the landscape is rich in archaeology, with the remains of enclosures, forts and burial chambers.

There are many other areas of heathland in Pembrokeshire, fragments of what was once an extensive network of commons. Formerly part of the agricultural system, these were used by local people for grazing, the cutting of gorse and the digging of clay, but many are now overgrown by scrub. The National Trust is working with local partners to restore these heaths by traditional methods of management, such as grazing and controlled burning.

Access

St David's Head is accessible by minor road from St David's. There is a car park at Whitesands Bay (charge), from which a network of footpaths crosses the headland and links up with the Pembrokeshire Coast Path. There is a café (not National Trust) at Whitesands Bay, and a National Trust visitor centre and shop in the centre of St David's. Some of the best examples of inland heath are the commons immediately adjacent to St David's, such as Waun Fawr, accessible off the B4583. Further information is available on 01348 837860.

When to visit

Spring and summer are best for the dramatic displays of plants. Birdlife is also most noticeable during the breeding season, with linnets, whitethroats and stonechats on the heathland and a variety of seabirds around the cliffs. Choughs, which feed on the areas of exposed turf, are also regular, and there is always a chance of seeing dolphins, porpoises and seals offshore.

County Down

Property Murlough National Nature Reserve

Murlough is an outstanding example of dune heath, with the oldest dunes having been dated to over 6,000 years ago and the highest reaching 36 metres (118 feet) or so. This is a nationally rare and declining habitat, and the reserve also includes areas of grassland and scrub, which together form a complex landscape of immense wildlife value. A wide range of plants occurs here, including two species of orchid, carline thistle, oysterplant, fern moonwort and devil's bit scabious, on which the larvae of the rare marsh fritillary feed. Invertebrates include many interesting dune creatures, such as the minotaur beetle and a large variety of solitary bees and wasps.

Access

Open access to most areas throughout the year; car park off the A24 near Slidderyford Bridge (charge for non-National Trust members). Information centre open seasonally. A network of footpaths leads across the dune heath to the shore and there are regular guided walks and other activities. Further information is available on 028 4375 1467. Dogs must be kept on leads at all times.

When to visit

Spring is especially delightful at Murlough. Birdlife is particularly noticeable at this time, and includes stonechat, linnet and excellent numbers of cuckoo, which can be seen patrolling the dunes looking for meadow pipit nests in which to lay their eggs. There is also a spectacular display of bluebells on the older fixed dunes at the northern end of the site.

Further reading

The following is a list of useful books and articles for those who wish to find out more about heathland. Many can be found on the shelves of public libraries, through which the more obscure and out-of-print titles can often be ordered.

An excellent account is Nigel Webb's *Heathlands*, The New Naturalist, Collins (1986). Also helpful and interesting are:
Lee Chadwick, *In Search of Heathland*, Dennis Dobson (1982)
Phil Colbourn and Bob Gibbons, *Britain's Natural Heritage*, Blandford Press (1987)
Oliver Rackham, *The Illustrated History of the Countryside*, Weidenfeld and Nicolson (1994)
Colin Tubbs, *The New Forest*, The New Naturalist, Collins (1986)
Thought-provoking ideas about the origins of heathland are explored in Frans Vera's *Grazing Ecology and Forest History*, CABI Publishing (2000)

Details on the ecology of heathlands can be found in:
C.H. Gimingham, *Ecology of Heathlands*, Chapman and Hall (1972)
C.H. Gimingham, *An Introduction to Heathland Ecology*, Oliver & Boyd (1975)
English Nature has published a wide range of material relating to heathland ecology and management; research reports nos 95, 101, 133, 170, 188, 271, 291 and 422 are especially relevant, as are English Nature Science series no.11 *The lowland heathland management booklet*, author Nick Michael (1996), and no.12 *Lowland heathland: the extent of habitat change*, author Lynne Farrell (1987, updated 1993)

Relatively little has been published that deals specifically with the social history of heathland, but the following titles are relevant and useful:
Rozanne Arnold, *Historical Heathland Survey of Corfe Castle Estate*, unpublished report for the National Trust (1999)
Eleanor Cooke, *Who Killed Prees Heath?*, Shropshire Wildlife Trust (1991)
Peter Francis, Jane Price and Kim Yapp, *Never On A Sunday*, Scenesetters/Shropshire Mines Trust (2000)
Chris Howkins, *Heathland Harvest* (1997) www.chrishowkins.com
Peter Kirby, *Forest Camera – A Portrait of Ashdown*, Sweethaws Press (1998)
A.T. Lucas, *Furze: a survey and history of its uses in Ireland*, Stationery Office, Dublin (1960)
Frances Mountford, *A Commoner's Cottage*, Alan Sutton Publishing (1992)

As far as monographs on individual heaths are concerned, the following is an excellent example and model:
J.J. Tucker, S. Zaluckyj & P.J. Alma, *Hartlebury Common: A Social and Natural History*, Hereford and Worcester County Council (1986)

Some of the most absorbing and evocative accounts of heathland were written many decades ago. Interesting descriptions of heathland's heyday appear in the following:
W.G. Clarke, *In Breckland Wilds*, Robert Scott (1925)
W.H. Hudson, *Hampshire Days*, Longman, Green & Co (1903), also Oxford University Press (1980)

Further historical context can be found in the novels of Thomas Hardy, notably in *The Return of the Native* (1878) and *The Withered Arm* (1888), and in S. Baring-Gould's *The Broom-Squire* (1913)

Early heathland conservation efforts are described in Norman Moore's fascinating *The Bird of Time*, Cambridge University Press (1987)

Excellent European context can be found in:
Kenneth Olwig, *Nature's Ideological Landscape*, George Allen & Unwin (1984)
Svein Haaland, *Fem tusen år med flammer,* Vigmostad Bjørke (2002)
Information about the Heathcult project can be found at www.nationaltrust.org.uk/heathlands/html/english/project/index.htm

Useful articles include:
John Box, 'Conservation or Greening?', *British Wildlife* Vol.4 (1993); 'Mineral Extraction and Heathland Restoration', *Mineral Planning* 80 (1999); 'Natural Legacies: Mineral Workings and Nature Conservation', *Mineral Planning* 68 (1996); 'Nature Conservation and Post-Industrial Landscapes', *Industrial Archaeology Review* XXI:2 (1999)
Lesley Haskins, 'Heathlands in an Urban Setting', *British Wildlife* Vol.11 No.4 (April 2000)
Nigel Webb, 'The traditional management of European heathlands', *Journal of Applied Ecology* 35 (1998)

National Trust Publications

The National Trust publishes a wide range of books that promote both its work and the great variety of properties in its care. In addition to more than 350 guidebooks on individual places to visit, there are currently over 70 other titles in print, covering subjects as diverse as gardening, costume, food history and the environment, as well as books for children. These are all available via our website **www.nationaltrust.org.uk** and through good bookshops worldwide, as well as in National Trust shops and by mail order on 01394 389950. The Trust also has an academic publishing programme, through which books are published on more specialised subjects such as specific conservation projects and the Trust's renowned collections of art.

Details of all National Trust publications are listed in our books catalogue, available from The National Trust, 36 Queen Anne's Gate, London SW1H 9AS – please enclose a stamped, self-addressed envelope.

Heathland is the third in a new National Trust series *Living Landscapes*. Appreciation of landscape dates back centuries, but a balanced understanding of the value of human interaction with the environment has only come about more recently. Through *Living Landscapes* we aim to explore this interaction, drawing on the vast range of habitats and landscapes in the Trust's care and on the bank of expertise the Trust has acquired in managing both these and the wildlife they support.

Each book in the series will explore the social and natural history of a different type of landscape or habitat. Beautifully illustrated with specially commissioned artwork and a range of stunning contemporary photographs and historical material, this series will appeal to all those with an interest in social history, wildlife and the environment. Further details are available on our website (see above).

Picture Credits

Page 4 David Woodfall/Woodfall Wild Images
Page 8 David Kjaer/rspb-images.com
Page 11 Ancient House Museum, Thetford
Page 12 James Parry
Page 13 Artista
Page 14 NTPL/Derek Croucher
Page 15 NTPL/Derek Croucher
Page 17 David Woodfall/Woodfall Wild Images
Page 20 Andy Hay/rspb-images.com
Page 21 Courtesy of Tomorrow's Heathland Heritage, adapted by Artista
Page 22 David Woodfall/Woodfall Wild Images
Page 23 NTPL/Derek Croucher
Page 24 David Woodfall/Woodfall Wild Images
Page 26 Roger Holman
Page 27 Rick Keymer Collection
Page 29 DNH&AS at Dorset County Museum
Page 30 Tate, London 2003
Page 31 NTPL/David Sellman
Page 32 DNH&AS at Dorset County Museum
Page 33 Peter Kirby
Page 35 Robin Bush/Nature Photographers
Page 36 Félix Arnandin © Parc Naturel Régional des Landes de Gascogne
Page 38 Hans Kampf, www.kampf.nl/hans
Page 39 Norfolk Museums and Archaeology Service /Derek A. Edwards
Page 41 NT
Page 43 British Library, Luttrell Psalter, 1000144.071
Page 45 British Library, Luttrell Psalter, 1000317.011
Page 48 Reproduced from *The Norfolk We Live In*, George Nobbs Publishing (1958)
Page 51 James Parry
Page 52 Alan Miller
Page 55 Ancient House Museum, Thetford
Page 56 Forest Enterprise
Page 58 Statens museum for Kunst, Copenhagen
Page 61 DNH&AS at Dorset County Museum
Page 62 The New Forest Ninth Centenary Trust
Page 65 Rick Keymer/Brenda Norrish
Page 66 James Parry
Page 67 © National Museums & Galleries of Wales
Page 68 Peter Kirby
Page 69 Verwood Historical Society
Page 71 NT/Andrew Tuddenham
Page 72 NT
Page 73 Ancient House Museum, Thetford
Page 74 Private Collection/Fiona Dickson
Page 76 James Parry
Page 77 Ancient House Museum, Thetford
Page 79 Richard Sorrell/Foundation for Art, NTPL
Page 80 Peter Kirby
Page 82 David Woodfall/Woodfall Wild Images
Page 85 Andy Hay/rspb-images.com
Page 89 Ernie James/rspb-images.com
Page 92 Robin Chittenden
Page 99 (top) Cambridge University Collection
Page 99 (bottom) Reproduced by permission of the Ordnance Survey of Northern Ireland on behalf of the Controller of Her Majesty's Stationery Office/30025 © Crown copyright 2003
Page 104 NTPL/Derek Croucher
Page 114 Roger Holman
Page 117 NTPL/Derek Croucher
Page 120 NT/Andrew Tuddenham
Page 123(top) Robin Bush
Page 123(bottom) Andy Fale
Page 124 NTPL/Derek Croucher
Page 125 Andy Fale
Page 126 Roger Tidman/Nature Photographers
Page 127 NT
Page 130 NT/Geoff Hann
Page 132 Paul Barwick and Forest Enterprise
Page 134 Hardy's Egdon Heath Project
Page 135 Arthur Nash
Page 137 NTPL/Derek Croucher
Page 139 NTPL/Joe Cornish
Page 141 NTPL/Ian West
Page 143 NTPL/John Miller
Page 145 David Woodfall/Woodfall Wild Images
Page 147 NTPL/Geoff Morgan
Page 149 NTPL/Joe Cornish

Inside artwork by John Davis
Scraperboard artwork by Alison Lang

Index

M

N

O

P

Q

R

S

T

U

V

W

Y

Z